Solange Batista Damasceno

Fauna diversity in the Miriti Environmental Protection Area, Amazonas, Brazil

Solange Batista Damasceno

Fauna diversity in the Miriti Environmental Protection Area, Amazonas, Brazil

Fauna diversity, Miriti River, Manacapuru, Amazonas, Brazil

ScienciaScripts

Imprint

Any brand names and product names mentioned in this book are subject to trademark, brand or patent protection and are trademarks or registered trademarks of their respective holders. The use of brand names, product names, common names, trade names, product descriptions etc. even without a particular marking in this work is in no way to be construed to mean that such names may be regarded as unrestricted in respect of trademark and brand protection legislation and could thus be used by anyone.

Cover image: www.ingimage.com

This book is a translation from the original published under ISBN 978-613-9-67951-5.

Publisher:
Sciencia Scripts
is a trademark of
Dodo Books Indian Ocean Ltd. and OmniScriptum S.R.L publishing group

120 High Road, East Finchley, London, N2 9ED, United Kingdom
Str. Armeneasca 28/1, office 1, Chisinau MD-2012, Republic of Moldova, Europe
Printed at: see last page
ISBN: 978-620-8-18832-0

CONTENTS

ACKNOWLEDGMENTS

To my parents **Severino Gerônimo Damasceno** and **Maria Neuza Batista Damasceno**, for their family upbringing, moral principles and religiosity.

To my children **Fabricio, Fernanda and Heloise**, who taught me the unconditional gift of love.

To my husband **Edimilson Barbosa Lima**, friend, companion and encourager of my academic growth.

To my **friends** who, in some way or another, were always there to give me a word of encouragement so that I wouldn't give up along the way.

And to my brothers and sisters by blood and heart, **Pedro, Severino Filho, Antônio Cezar, Mârio, José Alzenir, Neuzilane, Rozilane, Rosângela, Hermenegildo, Màrcio Célio**, who gave us support and encouragement at many moments during our academic life.

SUMMARY

The diversity of fauna catalogued in the Miriti Environmental Protection Area amounted to 163 species from 80 different families. Of the 163 species of fauna identified, 26 were from the Herpetofauna (16%), 59 from the Avifauna (36%), 24 from the Mastofauna (15%) and 54 from the Ichthyofauna (33%). And of the 80 families of fauna identified in the survey, 15 were from the Herpetofauna (19%), 28 from the Avifauna (35%), 19 from the Mastofauna (24%) and 18 from the Ichthyofauna (22%). Of the total fauna cataloged, twelve (12) species are on the Official List of Endangered Brazilian Fauna, according to MMA Ordinances No. 444/2014 and No. 445/2014: *Saimiri sciureus* (scent monkey), *Priodontes maximus* (armadillo canastra), *Leopardus pardalis* (maracajà cat), *Puma concolor* (brown jaguar), *Panthera onca* (jaguar), *Pteronura brasiliensis* (giant otter), *Crenicichla cincta* (jacundà), surubim, *Semaprochilodus insignis* (jaraqui-escama grossa), *Semaprochilodus taeniurus* (jaraqui-escama fina), *Colossoma macropomum* (tambaqui) and *Arapaima gigas* (pirarucu). We can conclude from the fieldwork that the diversity of fauna in the Miriti APA has uniform characteristics in its entirety. With regard to certain groups of fauna (avifauna, mastofauna and herpetofauna), we can say that there is a constancy in their identification along the entire route surveyed, but with regard to the ichthyofauna, it was observed that in the region upstream of the Miriti bathing resort there is a diversification of the species catalogued in this research, and therefore the greatest diversity of bottom fish (leather), called catfish. In summary, we can say that the biological diversity in the Miriti APA is in fact surviving and coexisting between the natural and environmental degradation as a result of the impacts that continue to advance, and that there is a need for energetic and structuring measures to be taken in this area, especially with regard to the management of water resources and all that surrounds it.

Keywords: Diversity; Fauna; APA; Catalog; Community; Miriti.

INTRODUCTION

The aim of this work is to research and investigate the diversity of fauna in the four major faunal groups: Avifauna, Hepertofauna, Ichthyofauna and Mastofauna in the Miriti Environmental Protection Area, in the municipality of Manacapuru, Amazonas, during the flooding period of the Solimoes River.

Given the need for knowledge of wild fauna in the Amazon, especially in Amazonas and more specifically in the areas of transition from rural to urban, such as the case of the Miriti APA, the subject of this research, it has a very high degree of importance due to the analysis of biodiversity, seen as a renewable natural resource, linked to conservation and sustainable use.

In this sense, the present study elucidates the cataloguing of Fauna Diversity in different areas of the Miriti APA, as well as the identification of specimens by fauna group in terms of family, species and popular name, statistically visualizing individuals by fauna group, classify the taxonomic groups according to ICMBIO's current List of Brazilian Fauna Species Threatened with Extinction, thus consolidating significant data that can serve as a guide for other research and for the management of the APA.

The research was divided into three phases: 1. Cataloging of primary data (fieldwork); 2. Data tabulation (identification of specimens by fauna group - family, species and popular name); statistics of individuals by fauna group; classification of taxonomic groups according to the current List of Brazilian Fauna Species Threatened with Extinction, MMA Ordinances No. 444/2014 and No. 445/2014; 3. Preparation of the research description (dissertation).

The work is sequentially distributed into chapters: Chapter 1: Biodiversity in the Historical Context, Chapter 2: Impacts on Biodiversity in Brazil, Chapter 3: Law No. 9.985/00 and the structure of the National System of Conservation Units, Chapter 4: Study Area, Chapter 5: Materials and Methods, Chapter 6: Results, Chapter 7: Discussions, Chapter 8: General Conclusions and Chapter 9: Recommendations.

General objective

To catalog the fauna diversity of the four major faunal groups, Avifauna (birds), Hepertofauna (reptiles and amphibians), Ichthyofauna (fish), and Mastofauna (mammals), in the Miriti Environmental Protection Area, Municipality of Manacapuru, Amazonas during the flooding period of the Solimoes River.

Specific objectives

-       Cataloging the fauna in different areas of the Miriti APA;

-       Identify the specimens by fauna group in terms of family, species and popular name;

-       Statistical analysis of individuals by fauna group;

-       Classify the taxonomic groups according to the MMA's current List of Brazilian Fauna Species Threatened with Extinction;

-       Consolidate the survey data.

THEORETICAL FIELD

CHAPTER 1: BIODIVERSITY IN THE HISTORICAL CONTEXT

Brazil stands out on the world stage for having a very significant biological diversity. It is one of the six countries with the greatest biodiversity on the planet, alongside Indonesia, Zaire, Colombia, Mexico and Peru, with an estimated two million species and approximately two hundred thousand described.

However, its biological wealth has been exploited in a disorderly and predatory manner since colonial times, which has contributed to putting some species at risk of extinction. According to the current Lists of Brazilian Fauna Species Threatened with Extinction (MMA Ordinances No. 444/2014 and No. 445/2014), there are 1,173 species.

Fauna plays a role in maintaining a healthy environment by providing the services needed to sustain human life, such as food, pollination and dispersal of plants, maintaining the balance of populations and pest control.

The more endangered species there are, the greater the risk suffered by the environment in which these species are found, i.e. there is a reduction in the quality of environmental resources such as water, fertile soil, air and products from nature.

The value of Biodiversity is incalculable and its reduction jeopardizes the sustainability of the environment, the availability of natural resources and thus life itself on Earth. Its conservation and sustainable use, on the other hand, result in incalculable benefits for humanity (Goeldi Museum, 2016).

In addition to the environmental services it provides, such as water purification, nutrient cycling and maintaining climatic conditions, biological diversity is an important source of resources for food, medicinal and industrial applications, among others. Brazil is home to around 20% of the world's biodiversity, mostly distributed in forest ecosystems. The Amazon rainforests account for around 25% of the planet's remaining forests, and in Brazil, they occupy almost half of the country's territory, and are of great strategic value to the country.

According to the Ministry of Science and Technology (MCT), the complex task of discovering, describing, characterizing and making good use of the products derived from Brazil's enormous biological diversity, as well as understanding patterns of change in the structure and function of biodiversity and their impact on society, requires a cooperative and articulated scientific effort, INPA (2008).

According to HNKU (1991), doing science means working in a group: it means taking on collective commitments. It is by definition "a collective activity, an activity organized in places and through institutions. In this sense, scientific practices are seen as dependent on human action and are related to social and cultural factors in the construction of scientific facts.

Given the great challenge of studying and researching the Amazon, especially biodiversity in the state of Amazonas, a number of institutions have been created which have become a reference over the last 100 years and which have contributed greatly to knowledge in the state. As well as participating in the implementation of major projects locally, and through studies mapping (cataloging) diversity and thus contributing to guiding strategic agendas for Amazonas.

According to Scopus, in the period from 1960 to 2013 six local teaching and research institutions had a high rate of publications linked to the research carried out, among them: Universidade Federal do Amazonas - UFAM, Instituto Nacional de Pesquisas da Amazônia - INPA, Universidade do Estado do Amazonas - UEA, Fundaçâo de Medicina Tropical Doutor Heitor Vieira Dourado - FMT-HVD, Instituto Federal de Educaçâo, Ciência e Tecnologia - IFAM and Fundaçâo Centro de Controle de Oncologia do Estado do Amazonas - FCECON (figure 1.1).

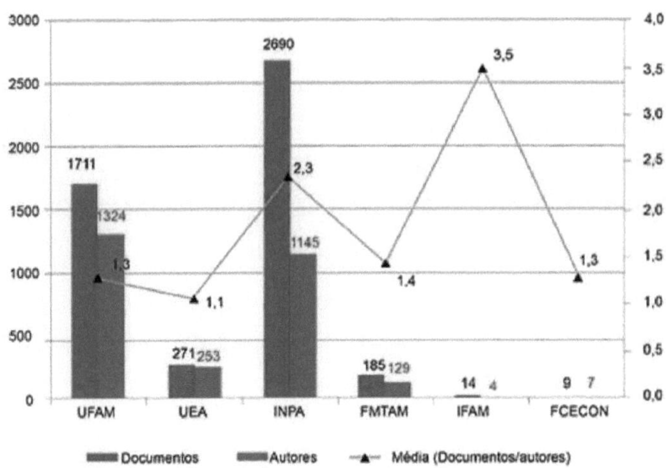

Figure 1.1: Total publications, authors and average for Amazonas, 1960 - 2013.

Source: SCOPUS.

The advance of science in the emerging economies is standing out in relation to the BRIC countries, which have been developing new institutional networks for scientific research, "putting in check" the hegemony of the economic and scientific superpowers established since 1945. Brazil has been standing out in this new scenario, especially in the last ten years, with important partnerships created with neighboring countries and expanding existing ones with traditional economies (USA, France and Germany).

With regard to scientific collaborations, the study shows that INPA, with important international collaborations, and UFAM, especially at the national level, have important partnerships with teaching and research institutions. These two institutions are historically important in the process of scientific development in the state.

In the case of younger institutions such as UEA, for example, the figure shows that these institutions (UEA, FMT-HVD, IFAM and FCECON) have fewer collaborations.

Among the three main local institutions (UFAM, INPA and UEA), INPA has the highest productivity in the state (2.3), followed by UFAM (1.3) and UEA, with (1.1) document per author.

We can say that regional research is very important and that much of the information on biological diversity in the Amazon region and in the state of Amazonas is the result of studies carried out by these institutions.

CHAPTER 2: IMPACTS ON BIODIVERSITY IN BRAZIL

The international scientific community, as well as governments and non-governmental environmental organizations, have been warning about the loss of biological diversity around the world, particularly in tropical regions. The biological degradation that is affecting the planet has its roots in the contemporary human condition, aggravated by the explosive growth of the human population and the unequal distribution of wealth. The loss of biological diversity involves social, economic, cultural and scientific aspects.

In recent years, human intervention in previously stable habitats has increased significantly, leading to greater losses of biodiversity. Biomes are being occupied at different scales and speeds and extensive areas of native vegetation have been devastated in the Cerrado of Central Brazil, the Caatinga and the Atlantic Rainforest and the Amazon.

According to MMA (2016), the main processes responsible for biodiversity loss are:

*   habitat loss and fragmentation;

*   introduction of exotic species and diseases;

*   overexploitation of plant and animal species;

*   use of hybrids and monocultures in agro-industry and reforestation programs;

- contamination of soil, water and atmosphere by pollutants; and

- climate change.

There are three main reasons for concern about the conservation of biological diversity. Firstly, because it is believed that biological diversity is one of the fundamental properties of nature, responsible for the balance and stability of ecosystems. Secondly, because it is believed that biological diversity represents an immense potential for economic use, especially through biotechnology. Thirdly, because it is believed that biological diversity is deteriorating, with an increase in the rate of species extinction, due to the impact of anthropogenic activities.

According to ISA (2016), species extinction, which is the disappearance of species from a given environment or ecosystem, is inherent to the biological process and concomitant with the emergence of life on earth. An example is the disappearance of the dinosaurs, which occurred millions of years ago, before the emergence of the human species. In other words, extinction is as natural an event as the emergence of new species, and occurs independently of human actions, due to natural disasters, competition for food, space and other resources and genetic mutations, among others.

The natural extinction of species, with the exception of those caused by natural disasters, is a slow event that takes thousands or even millions of years to occur. Human participation in the process of species extinction has led to an increase in the rate of extinction, making humanity the main force in triggering this process.

Through the overexploitation of species and environmental resources such as water, soil, minerals and the conversion of environments into more simplified production systems that are incapable of maintaining the biodiversity of habitats, species, processes and interactions, humanity has unleashed a cycle of species extinction that is unprecedented in the geological history of the earth.

The direct and indirect alterations caused by habitat conversion have led to an extremely fragmented landscape, dominated by agropastoral systems, abandoned areas of predatory and unsustainable extraction, the expansion of urban areas and an increase in the production and incorrect disposal of waste and the expansion of traffic vectors, such as the road, rail and river network.

These changes increase environmental fragility and the degree of isolation between natural populations, reducing gene flow, which can lead to a loss of genetic variability.

The introduction of invasive alien species is another important indirect alteration that is becoming a major threat to biodiversity.

Conserving Brazilian biodiversity and managing the conflict between conservation and predatory development is one of Brazil's biggest challenges today. The Ministry of the Environment, the central body of the National Environment System and whose mission is to promote the adoption of principles and strategies for the knowledge, protection and recovery of the environment, the sustainable use of resources and the inclusion of sustainable development in the formulation and implementation of public policies, has become responsible for a series of actions to guide this process:

1)   the drawing up of lists of endangered species, with the aim of quantifying the problem and allowing actions to be directed towards solving it, including the creation of restrictions on use;

2)   the development, dissemination and implementation of specific species protection and recovery policies;

3)   building a development model that ensures the sustainable use of biodiversity components.

By indicating the species threatened with extinction, the lists guide the application of other environmental laws, such as the aggravating penalties in the Environmental Crimes Law (Law no. 9.605/1998), to curb trafficking and illegal trade in species, in accordance with the annexes to the Convention on International Trade in Endangered Species of Wild Fauna and Flora (CITES) and to guide programs and action plans for the conservation and recovery of species of fauna and flora.

These lists are used in the definition of priority areas for biodiversity, in the establishment of new Conservation Units, in the definition of conservation guidelines and targets and measures to mitigate environmental impacts, in the licensing of projects, in access to genetic resources and in the management of fishing resources, in the management of forest resources, as well as in the application and orientation of funding for scientific research. Lists of endangered species are therefore an important public policy tool, which should be used wisely and sparingly, in favor of maintaining and recovering Brazil's rich biodiversity, supporting decision-making at local and global levels, ISA (2016).

At the international level, Brazil has ratified three conventions that provide the legal framework for the differentiated treatment of species considered to be threatened with extinction: the Convention for the Protection of the Flora, Fauna and Natural Scenic Beauties of the American Countries; the Washington Convention on International Trade in

Endangered Species of Wild Fauna and Flora (CITES) and the Convention on Biological Diversity (CBD).

At the national level, the Fauna Protection Law (No. 5,197/1967) establishes in its Article 1 that "animals of any species, at any stage of their development and living naturally outside captivity, constituting wild fauna, as well as their nests, shelters and natural breeding grounds are the property of the State, and their use, persecution, destruction, hunting or gathering is prohibited" (MMA, 2014).

2.1. Threatened Species and the Risk of Extinction of Brazilian Fauna

According to the Brazilian Fauna Extinction Risk Assessment, there are currently 1,173 species of endangered fauna, which make up MMA Ordinances No. 444/2014 and No. 445/2014. These lists were drawn up based on the process of Assessing the Risk of Extinction of Brazilian Fauna, where for each species there is information on the taxonomic classification, category of risk of extinction, criteria, bibliographic references and summary of the justification that indicated the risk of extinction (MMA, 2016).

In the Assessment of the Risk of Extinction of Brazilian Fauna, a diagnosis is made of the risk of extinction of the species, in which the main threats are identified and located, the important areas for the maintenance of the species and compatibility with anthropic activities. This survey supports the Ministry of the Environment's (MMA) review of the Official National List of Endangered Fauna Species. This is a cyclical process, with reassessments every five years, since knowledge about a given species and its conservation status changes over time. The first assessment cycle ended in 2014 by the Chico Mendes Institute, in which 12,254 taxa of Brazilian fauna were assessed. Currently, in 2016, the second evaluation cycle is underway.

2.2. Public policies aimed at protecting Brazil's endangered fauna

According to Planalto (2015), between 2003 and 2014, the federal government reinforced its actions to protect Brazilian fauna and flora, especially the group of species threatened with extinction. In December 2015, the federal government released the largest assessment of animals in the world, an inventory that cataloged 12,256 species of Brazilian fauna.

The study was carried out by the Chico Mendes Institute for Biodiversity Conservation (ICMBIO), an agency of the Ministry of the Environment, and also identified 170 species that are no longer on the National List of Endangered Species, such as the humpback whale and the hyacinth macaw, both of which have populations that are recovering (figure 2.2.1).

Of the species catalogued, 4 are amphibians, 23 birds, 14 mammals, 2 reptiles, 45 terrestrial invertebrates and 82 fish and aquatic invertebrates.

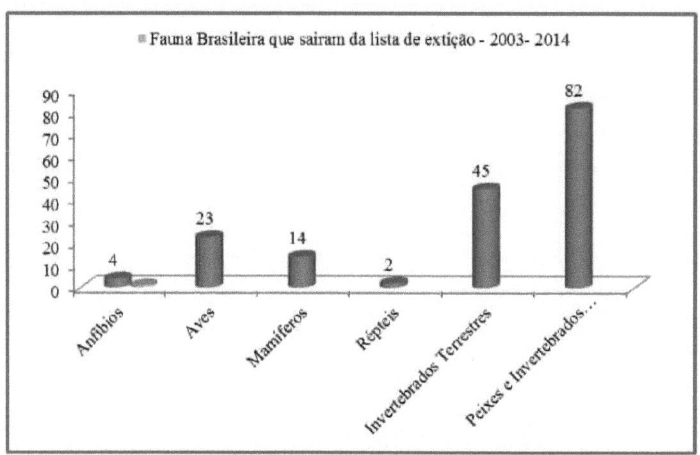

Figura 2.1: Species of Brazilian fauna that have been removed from the extinction list.

Source: Portal Planalto - Presidency of the Republic (2015).

The mapping of species of fauna and flora and of fish and aquatic invertebrates increased the number of species assessed by 800% compared to the last survey, published in 2003, as shown in figure 2.2 fauna in 2003, which totaled 1,137 to 12,256 species.

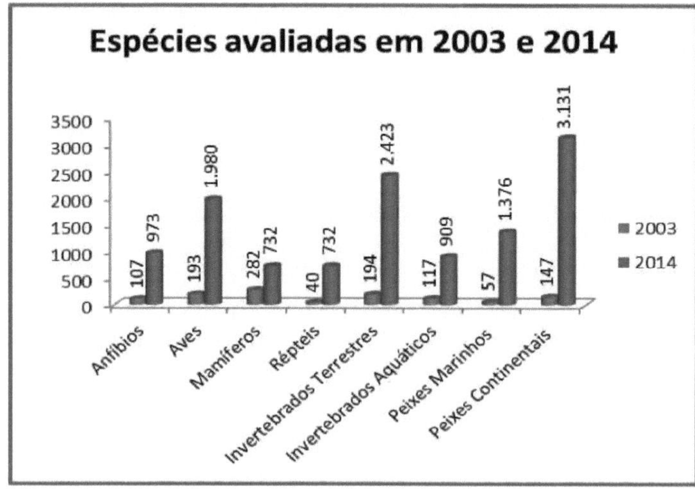

Figura 2.2: Brazilian Fauna Species evaluated between 2003 and 2014.

Source: Portal Planalto - Presidency of the Republic (2015).

According to Izabella Teixeira, Minister of the Environment in 2015, the study will help to create new conservation units, which in her opinion values integral protection in the country.

Since the National Program for the Conservation of Endangered Species, the institutional stance has become a benchmark because, in addition to complying with international risk assessment standards, with annual updates by species group and a general check every five years, Ordinance No. 43 of January 31, 2014, which established Pro-Species, also provides for National Action Plans (PANS) for the conservation of threatened animal groups. In the last ten years, the Chico Mendes Institute has already drawn up more than 50 NAPs, covering species such as the maned wolf, jaguar and parrots of the Atlantic Forest.

The criteria used to assess endangered species were the same as those adopted by the International Union for Conservation of Nature (IUCN) and under the terms of the Convention on Biological Diversity (CBD), ICMBIO (2014).

CHAPTER 3: LAW NO. 9.985/00 AND THE STRUCTURE OF THE NATIONAL SYSTEM OF CONSERVATION UNITS

The creation of conservation units is one of the duties of the state. According to Giansanti (1998), Conservation Units (CUs) are areas with natural characteristics of relevant value, legally protected and maintained under special management regimes.

These are representative natural areas set up for preservation, research, environmental education and leisure. Basically, the objectives of conservation units are to preserve biodiversity, protect rare, endemic or endangered species, preserve ecosystems, encourage the sustainable use of resources and protect natural or slightly altered landscapes, Giansanti (1998).

Thus, from the point of view of sustainable development, the existence of these units can stimulate and advance practices of sustainable use of resources, as in the case of National Forests, and combine the knowledge accumulated by traditional communities and modern science.

According to Magalhâes (2001), the so-called Conservation Units are territorial spaces, mostly made up of continuous areas where the aim is to preserve fauna, flora and natural beauty, as well as the environment as a whole, with the aim of perpetuating these spaces.

The International Union for Conservation of Nature (IUCN) defines Conservation Units as "areas defined by the Public Authorities for the protection and preservation of ecosystems in their natural and primitive state, where natural resources can be used indirectly without consumption", Magalhâes (2001).

As established in Law No. 9.985/00, a Conservation Unit (Art. 2, I) is a territorial space and its environmental resources, including jurisdictional waters, with relevant natural characteristics, legally established by the Public Power, with conservation objectives and defined limits, under a special administration regime, to which adequate protection guarantees are applied. It also instituted the so-called National System of Nature Conservation Units - SNUC and specifies the objectives of these protected spaces. They are:

I)   contribute to the maintenance of biological diversity and genetic resources in the national territory and jurisdictional waters;

II)   protecting endangered species at regional and national level;

III)   contribute to the preservation and restoration of the diversity of natural ecosystems;

IV)   promoting sustainable development based on natural resources;

V)   promote the use of nature conservation principles and practices in the development process;

VI)   protect natural and little-altered landscapes of outstanding scenic beauty;

VII)   protect relevant geological, geomorphological, speleological, archaeological, paleontological and cultural features;

VIII)   protecting and restoring water and edaphic resources;

IX)   recovering or restoring degraded ecosystems;

X)   provide means and incentives for scientific research activities, studies and environmental monitoring;

XI)   give economic and social value to biological diversity;

XII)   favor conditions and promote environmental education and interpretation, recreation in contact with nature and ecological tourism;

XIII)   protecting the natural resources necessary for the subsistence of traditional populations, respecting and valuing their knowledge and culture and promoting them socially and economically.

Traditionally in Brazil there has been a range of legislation aimed at protecting forests, but they were not considered to be environmental protection because they were based on economic interests, but in any case the environment was favored, Magalhaes (2001).

From the beginning of the 19th century, with the arrival of the royal family in Brazil,

Conservation Units were created and the creation of environmental protection spaces evolved, such as the Botanical Garden of Rio de Janeiro, created on March 10, 1811, with an area of 2,160 hectares, but currently with only 137 hectares.

3.1. Environmental Protection Units

Today there is a great diversity of Conservation Units in Brazil. There are various ways of grouping these representative areas of ecosystems, taking into account their objectives. Thus, some are totally untouchable, others allow indirect use and some allow rational economic exploitation.

According to Magalhâes (2001), the term Conservation Unit is the most appropriate generic name for Environmental Protection Units because it encompasses both preservation and conservation units. Thus, according to the aforementioned author, there would be two Environmental Protection Units: Preservation Units and Conservation Units.

The following belong to the Preservation Units: Forest Parks, Biological Reserves, Ecological Stations, Forest Gardens, Park Roads, Natural Monument, Sanctuary or Wildlife Refuge, Scenic River, Resource Reserve, Indian Reserve, Biosphere Reserve, World Heritage Reserve.

Conservation Units include: National, State and Municipal Forests, Environmental Protection Areas, Areas of Relevant Ecological Interest, Areas of Special Interest, Areas of Tourist Interest, Sites of Tourist Interest and Natural Areas. This systematization is quite logical, as it is consistent with the Forest Code.

According to Magalhães (2001), Preservation Units are those for indirect use, since they are intended for the conservation of biodiversity, scientific research, environmental education and recreation. Conservation Units, on the other hand, allow for the direct use of their resources, since as well as being intended for the conservation of biodiversity, they also allow for the sustainable use of natural resources, establishing development models. Therefore, the term Conservation Units does not lend itself to encompassing all forms of these environmental protection spaces. Thus, Environmental Protection Unit seems more appropriate because it is more comprehensive.

With the advent of Law No. 9.985/00, Conservation Units now form two groups (Art. 7): Integral Protection Units and Sustainable Use Units. The basic objective of the former is to preserve nature, with only indirect use of its natural resources permitted; the latter is to make nature conservation compatible with the sustainable use of part of its natural resources.

Law No. 9.985/00 considers indirect use to be that which does not involve the consumption, collection, damage or destruction of natural resources (Art. 2, IX); and sustainable use to be the exploitation of the environment in such a way as to guarantee the continuity of renewable environmental resources and ecological processes, maintaining biodiversity and other ecological attributes, in a socially just and economically viable manner (Art. 2, XI). This law seeks to standardize the nomenclature of Conservation Units by determining that existing categories should be re-evaluated within two years, with the aim of defining their destination based on the category and function for which they were created, in accordance with the provisions of its text.

According to Feldmann *apud* Giansanti (1998), the attributions of the different conservation units, according to current legislation, are as follows:

<u>National Parks</u> - These are areas of considerable size, defined by the Forest Code of 1965, created by the government for the purpose of ecological preservation and the protection of rare species, water resources and geological structures. They are intended for leisure and recreation, research and environmental education, and any possibility of exploitation or extraction of resources is prohibited.

<u>Biological Reserves</u> - These are characterized by containing ecosystems or specimens of fauna and flora of significant biological importance. They are areas of varying extent, created on public land, closed to visitors (with the exception of authorized groups of researchers) and with total restrictions on any form of resource exploitation.

<u>Ecological Stations</u> - These are areas of significant portions of natural ecosystems, created by federal law in 1981, intended for research, protection and environmental education. At least 90% of their area must be used for integral preservation, with the rest being used for research and education activities.

<u>National Forests</u> - Also created by the Forest Code of 1965, they are located mainly in the north of the country. They are large areas covered by native forests. They are intended for the sustainable use of wood and other forest products, the protection of water resources, wildlife management, leisure and recreation.

<u>Environmental Protection Areas (APAs)</u> - APAs are conservation units created to conserve wildlife, natural resources and gene banks, as well as preserving the quality of life of local inhabitants. They involve areas with a density of occupation, depending on environmental zoning and the participation of the population to achieve their objectives. They can be at federal or state level.

<u>Extractive Reserves</u> - Provided for by Law No. 7.804/89, extractive reserves are Union areas

used by concession, under the regulation of the federal and state governments.

There are no individual property titles. In these areas, traditional groups and cultures are dedicated to extracting products of commercial value, such as latex, Brazil nuts and vegetable oils, as well as non-predatory hunting and fishing and subsistence farming.

<u>Indigenous Lands</u> - There are 510 indigenous territories, covering around 891,000 km^2 (10.52% of the national territory). Of this total, only 255 have been demarcated. In addition to these territories, there are 18 reserves and 4 indigenous parks. The official population estimates and areas of occupation do not take into account the groups that live close to urban centers or that are not settled, Barreto (2009).

<u>Other conservation units</u> - In addition to the types of conservation units mentioned above, there are others involving both public and private land, such as areas of relevant ecological interest, areas of special protection and listed areas. The latter are provided for in the legislation that has regulated the process of landmarking since the 25th of 1937, covering areas whose conservation is of interest because of their historical, archaeological, environmental, geological, landscape or tourist value.

There are also what are known as production units, such as experimental stations and state forests, whose priority objectives are both experimentation and research and the production and marketing of forest products.

Consider also the existence of legal instruments such as EIA/RIMAS (Environmental Impact Studies - Environmental Impact Reports), which forecast environmental impacts and indicate occupation limits Giansanti (1998).

EMPIRICAL FIELD

CHAPTER 4: STUDY AREA

4.1. Location and context of the municipality of Manacapuru

The Miriti APA belongs geographically to the municipality of Manacapuru, which is located on the left bank of the Solimoes River. Manacapuru is the third largest municipality in the state of Amazonas. Known as the Little Princess of Solimoes, its main characteristics are the hospitality of its people and its natural beauty, including the Miriti Environmental Protection Area and the Piranha Sustainable Development Reserve.

4.1.1. Origin of the name Manacapuru

Manacapuru is a word of indigenous origin derived from the expressions Manacà and Puru. Manacà (*Brunfelsia hospeana*) is a Brazilian dicotyledonous plant from the Solanaceae family which means <u>Flower </u>in Tupi. Puru, from the same origin, means <u>adorned, tinted</u>. Therefore, Manacapuru in the indigenous Tupi language means "Tinted Flower".

4.1.2. History of Manacapuru

The city of Manacapuru, founded on February 15, 1978, originated from a village of Mura Indians, which was pacified in 1785.

At that time, on the banks of the Solimoes River, just below the mouth of the Manacapuru, there was a fishing factory called Caldeirâo, whose production supplied the military garrison based in Barcelos, the seat of the Captaincy.

In 1894, by State Law No. 83, Manacapuru was elevated to the category of village and the municipality was created, separated from the municipality of Manaus.

4.1.3. Administrative formation of the municipality of Manacapuru

Law No. 148, of August 12, 1865, created the parish of Nossa Senhora de Nazaré de Manacapuru, with its seat in the village of Manacapuru. Raised to the category of town, with the name of Manacapuru, by State Law No. 83, of 27.09.1894, it was dismembered from the municipality of Manaus. Installed on 16.06.1895.

The District of Manacapuru was created by Law No. 354 of September 10, 1901. In the administrative division for 1911, the municipality is made up of 13 districts: Manacapuru, Aiapuà, Arara, Beruri, Caapiranga, Campina, Conceiçâo de Manacapuru, Guajaratuba, Jaitenga, Manaquiri, Mundurucus, Paratari and Tamanduà.

In the tables of the general census of 1-IX-1920, the municipality is made up of 5 districts:

Manacapuru, Aiapuà, Campinas, Manaquiri, Terra Preta.

By virtue of Law No. 1,126, of November 5, 1921, the Manacapuru district was abolished and re-established the following year, according to Law No. 1133, of February 7.

Raised to the status of a city, with the name of Manacapuru, by State Act No. 1.639, of 16.07.1932.

By State Decree-Law No. 176, of 01.12.1938, the districts of Caapiranga and Beruri were created and annexed to the municipality of Manacapuru.

In the table established for the period 1939 to 1943, the municipality is made up of 3 districts: Manacapuru, Beruri and Caapiranga. This is how it remained in the territorial division dated 1-VII-1955.

By Municipal Law No. 14, of 07.06.1957, the district of Vila Rica was created and annexed to the municipality of Manacapuru. According to the administrative division in force in December 1959, three districts make up the municipality: Manacapuru, Beruri and Caapinanga.

In a territorial division dated 1-VII-1960, the municipality is made up of 4 districts: Manacapuru, Beruri, Caapiranga and Vila Rica.

By State Law No. 1, of 12.04.1961, the district of Beruri is separated from the municipality of Manacapuru and elevated to the category of municipality.

By State Law No. 7, of 09.04.1963, the districts of Caapiranga and Vila Rica were separated from the municipality of Manacapuru and elevated to the category of municipality.

In a territorial division dated 31-XII-1968, the municipality is made up of the headquarters district. It remains so in the territorial division dated 2014, IBGE (2015).

4.1.4 *Comparison of the area of the territorial unit between the municipalities surrounding Manaus - 2015*

In comparison with six municipalities around Manaus, Novo Airao, Presidente Figueiredo, Itacoatiara, Manacapuru, Rio Preto da Eva and Iranduba. Manacapuru is the 5th largest municipality in terms of land area in 2015, with 7,330.074 km^2 (figure 4.1). It had a population of 85,141 in 2010 and an estimated population of 95,330 in 2016, with a demographic density of 22.62 inhabitants per square kilometer in 2010.

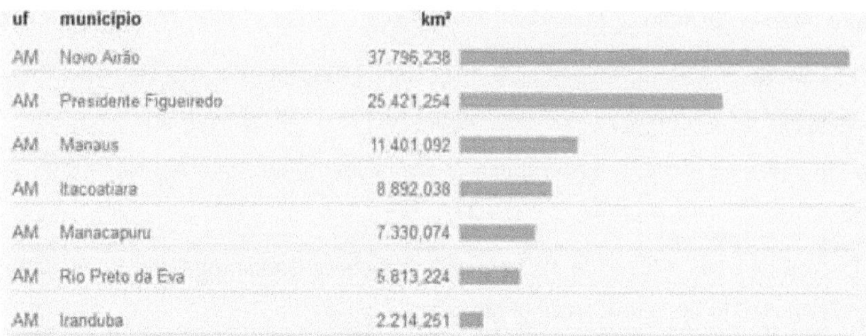

Figure 4.1: Comparison of the Municipalities Surrounding Manaus. **Source:** IBGE - Brazilian Institute of Geography and Statistics (2010).

4.1.5 Developments in the municipality in the last decade

Manacapuru's Municipal Human Development Index (MHDI) has increased considerably between 1991 and 2010, rising from 0.339 in 1991 to 0.437 in 2000 and reaching 0.614 in 2010. This

According to the UN (2014), several factors have contributed to the increase in life expectancy in Brazil. Between 1996 and 2006, the country halved its child malnutrition rate, from 13.4% to 6.7%. Actions to combat malnutrition include expanding the supply of doses of vitamin A and ferrous sulphate, as well as improving nutritional surveillance in municipalities with a malnutrition rate of more than 10%.

Based on this information at the national level, we understand that the municipality of Manacapuru has been experiencing positive progress in several aspects over the last decade, including the benefits received from the implementation of the Coari-Manaus Gas Pipeline Program, the construction of the bridge over the Negro River, which connects Manaus to the surrounding municipalities and currently the duplication of the AM 070 highway. These are factors that have influenced the increase in income of the residents of Manacapuru and consequently their quality of life has also suffered positive impacts.

4.2. Miriti Environmental Protection Area

The Miriti APA was created by Manacapuru City Hall in 1997, when the first steps were taken towards the creation of the Municipal Conservation Unit System - SMUC, which includes the Miriti Environmental Protection Area (APA) and the Piranha Sustainable Development Reserve - RDS Piranha. At the time of their creation in 1997, these two protected areas covered approximately 14.3% of the municipality's territory. At the time of their creation, approximately 600 families lived there, with a population of around 3,000 inhabitants, who depend directly and indirectly on the natural resources in these areas both for their survival and for their economies, PMM (1997).

The Miriti APA was created by Municipal Decree No. 009/97, and for the creation of this Unit there was no preliminary environmental study, nor was there any mapping of its geographical delimitation, which made research work in the area very difficult because we didn't know exactly where the APA began or ended.

The research was carried out in part of the APA and delimited through the technical methods and methodologies used for environmental studies, used for the purposes of obtaining data, cataloguing fauna and obtaining a significant sample, which led us to the objectives achieved (figure 4.2).

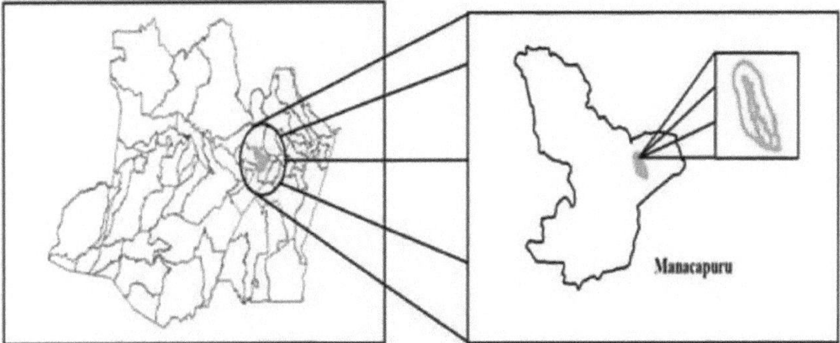

Figure 4.2: Research Area - Miriti APA. **Source**: Map adapted from the Digital Vector Base of the Institute of Geography and Statistics - IBGE and field research.

In fact, the APA of Miriti, the object of the research, was the Miriti basin and its surroundings and the two ecosystems, downstream and upstream of the Balneàrio Miriti, which are apparently identical, but have very different environmental characteristics when it comes to the diversity of local fauna.

4.2.1. APA downstream of the Miriti River

Downstream of the Miriti bathing resort, at the mouth of the Miriti River where it meets the Solimoes River, the Solimoes advances over the Miriti during the state's seasonal flooding and floods the flat or lowland areas (figure 4.3; a, b, c, d), causing great damage and discomfort to the community living in the outlying areas.

During the flood season, small and medium-scale artisanal fishing takes place in the area (figure 4.3; e) due to the phenomenon of piracema ("the *migratory movement of fish towards river sources for the purpose of reproduction"),* one of the reasons that probably impacts gene exchange in the area.

In the area downstream of Balneário do Miriti are the two water catchment points for the public supply of the city of Manacapuru, businesses of all kinds, as well as various leisure and tourism areas in the municipality (figure 4.3; f, g, h) including indigenous community housing.

Figure 4.3: Area of the Miriti APA comprising the basin as the object of research. **a)** Near Paraiso D'Angelo. **b)** Area where the Solimoes River flows into the Miriti River. **c)** and **d)** Houses flooded during the flood - Bairro de Sao Francisco. **e)** Small and medium-scale artisanal fishing. **f)** Miriti River in front of Mundo Amazônico. **g)** Miriti River - Water collection point. **h)** Miriti waterfront - Bairro da Liberdade. **Source**: Technical field collection - Solange B. Damasceno (2013-2015).

4.2.2. *APA upstream of the Miriti River*

The area upstream of the Miriti resort is considered to have calmer waters, as they are dammed by the construction of the bridge that forms part of the AM 070 highway (figure 4.4; c, d).

In this part you can also find a number of restaurants, of various constructions, some masonry or mixed, others floating, (figure 4.4; a, b, d) as well as a variety of leisure areas, including occasional ones, including clubs.

You can also find the infrastructure of a school, a community center and several residents in the area where the research data was catalogued.

Therefore, it is worth noting that the area upstream of Balneàrio Miriti is much better preserved, with characteristics of a primary area throughout, as shown in figure 4.4 (e, f, g, h).

Figure 4.4: Area upstream of Balneàrio Miriti. **a)** Bar Tia Alzira - near the bridge. **b)** Other bars upstream of the bridge. **c)** Road 070 crossing the Miriti River. **d)** Artisanal fishing and Repolho Restaurant. **e)** Residents of the Agua Preta creek. **f)** Along the Miriti River upstream

of the resort. **g)** Area at the entrance to the Agua branca creek. **h)** Area of igapó near the headwaters. **Source**: Technical field collection - Solange B. Damasceno (2013-2015).

When the Amazon rivers flood, the Miriti basin receives a much larger volume of water from the Solimoes river, which floods a large part of the basin, influencing almost all of its length (figure 4.5).

Figure 4.5: Miriti waterfront, Liberdade neighborhood. **Source**: PRF. Tourism - 2015

According to the observations made during the research, we can consider that the area studied can be characterized as a mixed area, where part is considered to be flooded forest or Alluvial Ombrophilous Dense Forest, Veloso *et al.* (1991). In the Amazon region, the Ombrophilous Dense Alluvial Forest receives the popular name of vàrzea or igapó depending on the color of the river water, designations adapted for scientific literature as being forests flooded by muddy water (vàrzea) or black/transparent water (igapó), Pires (1974).

Thus, as the basin is influenced by rivers with black water and also by muddy water, especially the Solimoes River during the flood season, it can be classified as dense alluvial forest (flooded and igapó), with a strong Amazonian lowland character.

In addition to the plant typology, a strong anthropogenic influence was found throughout the basin, since it is surrounded by farms on the one hand and the peri-urban area of the Municipality of Manacapuru on the other. It is also directly influenced by the Manoel Urbano highway, AM 070 and, consequently, by the Miriti resort and other resorts along its route, causing small, medium and large environmental impacts.

In all the areas where the data was collected, the composition of the environments was observed to check for animal specimens and the vast majority of the environments are uncharacterized or in a transitional or anthropized phase.

CHAPTER 5: MATERIALS AND METHODS

5.1. Materials

The materials used for the fieldwork were:

- Agenda for notes (various);

- Fauna cataloguing table;

- 50m tape measure;

- GPS;

- Clipboard;

- Camera;

- Làpis;

- Pen;

- Personal protective equipment (boots, gloves, blouse and long pants, others);

- Cloth-brimmed hat;

- Canteen for storing water;

- Repellent;

- Sunscreen;

- Dehydrated snacks (cereal bar, banana, others).

The materials used to process the data and prepare the dissertation were:

- Computer;

- Bibliographic, physical and digital collection;

- Laser printer;

- A4 paper.

5.2. Methodology

The methodologies used for cataloging the specimens of wild fauna in the field research were diverse. For Avifauna, Herpetofauna and Mastofauna, 26 terrestrial sampling sites (transects) were used (figure 5.1) along the APA, so specimens were cataloged in the lowland, slope and plateau areas. For the Ichthyofauna, 12 points were adopted for data collection, distributed along the Miriti River and at each point 4 different mm meshes were placed, as shown in figure 5.2.

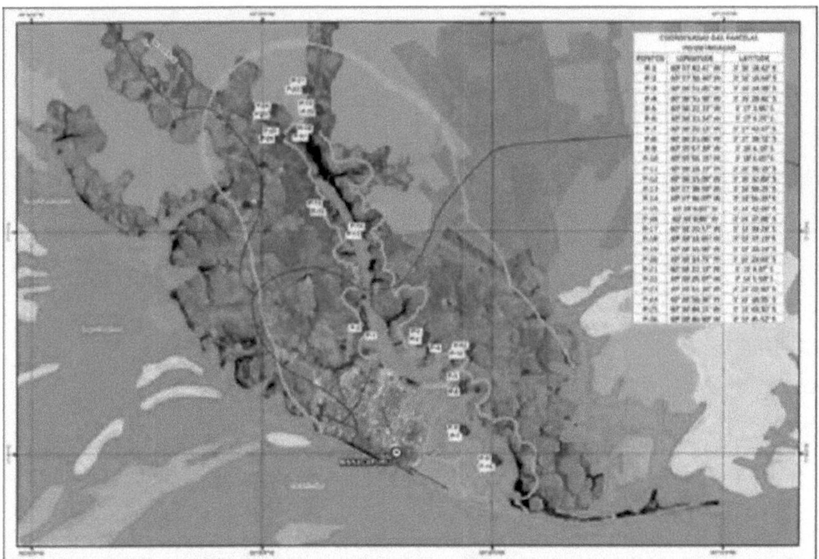

Figura 5.1: Distribution of the terrestrial sampling sites used to catalog Avifauna, Herpetofauna and Mastofauna. **Source**: Map based on the Digital Vector Base of the Institute of Geography and Statistics - IBGE and field research.

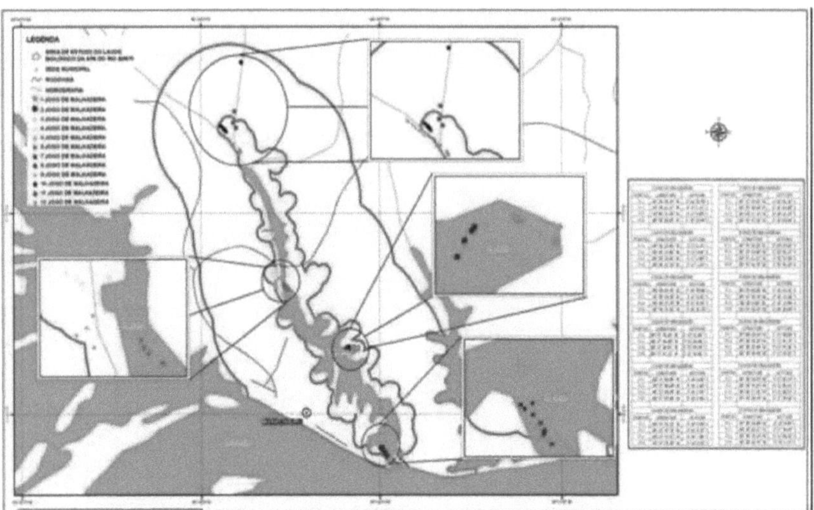

Figura 5.2: Distribution of Ichthyofauna data collection points along the Miriti River. **Source**: Map based on the Digital Vector Base of the Institute of Geography and Statistics - IBGE and field research.

The size of the terrestrial sampling sites, the standard used to inventory the fauna in the area that was the subject of the research, was based on the Terms of Reference (TR) of IPAAM - the Amazonas State Environmental Protection Institute (**Annex 1**), used in floristic inventories for the purposes of obtaining a single environmental license to suppress vegetation.

The TR was used because there is no specific delimitation for the inventories and/or cataloguing of fauna that has been instituted so far.

The measurement determined in the TR is 20m x 125m, totaling 2,500 m^2 per transect, as shown in figures 5.3 and 5.4, containing the geographic coordinates at the sampling vertices, tables 5.1 listing the Avifauna, Herpetofauna and Mastofauna and 5.2 listing the Ichthyofauna.

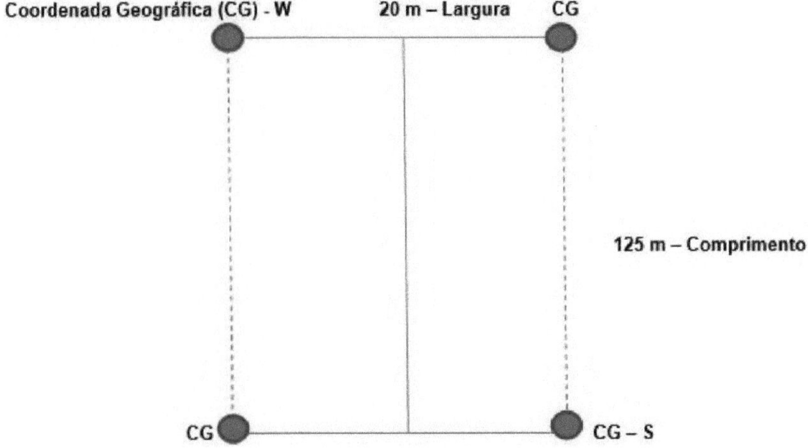

Figura 5.3: Standard sample site for cataloging wildlife. **Source**: Training Course on Wildlife Inventories handout - Solange B. Damasceno technical collection (2016). Damasceno (2016).

During the survey, it was necessary to adapt the pattern of the terrestrial transects into L, V or Z shapes according to the diagram in figure 5.4 and not necessarily all of them had a single direction, i.e. the size of the sampling area was maintained in the pattern mentioned above, but the direction of the transects was according to the topography of the terrain that was available to us in the field, thus preserving the total area inventoried per transect.

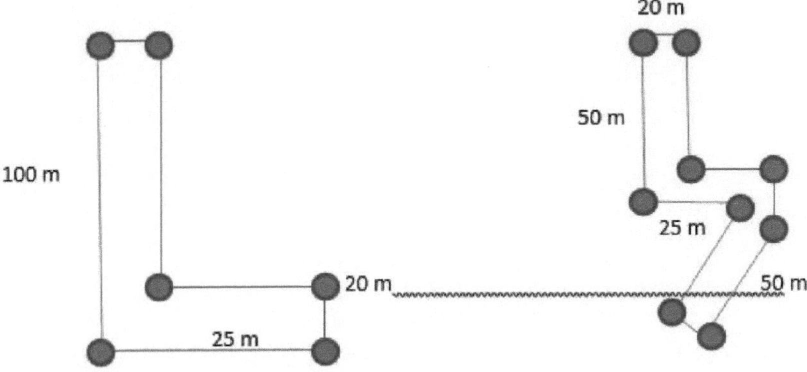

Figura 5.4: Direction of sampling sites (transects) adapted during wildlife cataloging. **Source**: Training Course on Wildlife Inventories handout - Solange B. Damasceno technical collection (2016). Damasceno (2016).

The ground transects were opened with the help of a field hand (figure 5.5), keeping a safe distance at all times, as there are many lianas and palm trees with thorny stems in the region and we were prone to accidents during the work.

In addition, we also noticed a great diversity of bees and wasps (marimbondos), so we had to be extra careful. Another fundamental factor was to watch out for venomous animals (snakes, scorpions and spiders), of which there is an abundant variety in the area.

Figura 5.5: Opening of terrestrial transects. **a)** Terrestrial transect on plateau. **b)** Terrestrial transect on plateau. c) Terrestrial transect on slope. **d)** Transect on riverbank in lowland area. **Source:** Technical field collection - Solange B. Damasceno (2013-2015).

Points (Transects)	S	W	Points (Transects)	S	W
	Transect			Transect	
P1	03°16'18.42"	60°37'42.41"	P14	03°14'56.25"	60°37'36.07"
P2	03°16'18.54"	60°37'38.46"	P15	03°14'42.09"	60°38'6.65"
P3	03°16'24.38"	60°36'51.05"	P16	03°14'37.86"	60°38'8.86"
P4	03°16'24.42"	60°36'51.98"	P17	03°13'39.29"	60°38'20.57"
P5	03°17'3.95"	60°36'22.23"	P18	03°13'37.13"	60°38'16.85"

Points (Transects)	S	W	Points (Transects)	S	W
P6	03°17'8.25"	60°36'21.34"	P19	03°13'20.19"	60°38'15.96"
P7	03°17'42.97"	60°36'20.13"	P20	03°13'23.63"	60°38'14.75"
Points (Transects)	**S**	**W**	**Points (Transects)**	**S**	**W**
Transect				**Transect**	
P8	03°17'39.72"	60°36'21.86"	P21	03°13'4.97"	60°38'22.19"
P9	03°18'4.18"	60°35'57.39"	P22	03°13'5.50"	60°38'25.97"
P10	03°18'6.65"	60°35'55.15"	P23	03°13'22.60"	60°38'51.36"
P11	03°16'36.15"	60°36'18.19"	P24	03°13'18.95"	60°38'50.36"
P12	03°16'32.89"	60°36'15.09"	P25	03°13'43.30"	60°38'44.15"
P13	03°14'59.25"	60°37'38.59"	P26	03°13'45.52"	60°38'46.90"

Table 5.1: Terrestrial Transects (Sample Sites) - Catalogue of Avifauna, Herpetofauna and Mastofauna.

The ichthyofauna data collection points for this research, both downstream and upstream of the Miriti spa, were in different environments, such as: in an area close to the cattle ranch within the APA, in an area of flooded forest according to the seasonal phenomenon of the region, in forest flooded all year round, called igapó forest, in an area with abundant aquatic macrophytes and in a flooded area characterized as lowland or floodplain, close to the Miriti bathing resort (figure 5.6).

Figure 5.6: Ichthyofauna data collection points. **a)** and **b)** Ichthyofauna collection point downstream of the Miriti bathhouse. **c)** Ichthyofauna collection point in an area near a farm

within the APA downstream of the Miriti bathhouse. **d)** Ichthyofauna collection point in a flooded forest area downstream of the Miriti bathhouse. **e)** Ichthyofauna collection point in flooded forest upstream of the Miriti resort. **f)** Ichthyofauna collection point in igapó forest upstream of the Miriti resort. **g)** Ichthyofauna collection point in an area with aquatic macrophytes upstream of the Miriti resort. **h)** Ichthyofauna collection point in an annually flooded area near the Miriti resort. **Source:** Technical field collection - Solange B. Damasceno (2013-2015).

In the case of cataloguing the ichthyofauna, 12 sets of meshes with 4 different millimetre meshes (25, 35, 40 and 50mm) were used, with one (01) 25mm mesh, one (01) 35mm mesh, one (01) 40mm mesh and one (01) 50mm mesh for each set (table 5.2).

The meshes were placed in different locations, upstream and downstream of the Miriti resort, for a period of 24 sequential hours, which were checked every two hours to remove the fish caught in the meshes and identify the specimens.

The quantity of fishing gear (mesh) used in the environments was based on the sample size delimited for the research. The type of mesh (mm) used *in* the research was based on observation *in loco on* the Miriti River and informal conversations with fishermen from the Miriti resort area, who told us which meshes they used to fish in the Miriti River.

Points	S	W	Mesh	Knitting machines
1	03°16'59.94"	60°36'26.34"	25	
2	03°17'2.10"	60°36'22.32"	35	1st Mallet Game
3	03°17'0.60"	60°36'21.66"	40	
4	03°16'59.64"	60°36'26.82"	50	
5	03°17'2.10"	60°36'33.06"	25	
6	03°17'0.66"	60°36'31.80"	35	2nd Mallet Game
7	03°17'3.84"	60°36'33.90"	40	
8	03°17'1.02"	60°36'32.16"	50	
Points	S	W	Mesh	Knitting machines
9	03°16'58.86"	60°36'29.34"	25	
10	03°17'9.00"	60°36'29.94"	35	3rd set of knitting machines
11	03°17'8.70"	60°36'31.14"	40	
12	03°16'59.76"	60°36'32.40"	50	
13	03°15'55.86"	60°37'45.66"	25	
14	03°15'54.96"	60°37'46.68"	35	4th Knitting game
15	03°15'54.90"	60°37'48.42"	40	

16	03°15'54.90"	60°37'41.22"	50	
17	03°16'0.60"	60°37'43.86"	25	
18	03°16'3.60"	60°37'46.68"	35	5th Knitting game
19	03°16'2.46"	60°37'44.94"	40	
20	03°16'5.46"	60°37'43.68"	50	
21	03°16'5.46"	60°37'35.22"	25	
22	03°16'4.26"	60°37'35.76"	35	6th Knitting game
23	03°16'7.86"	60°37'31.92"	40	
24	03°16'6.36"	60°37'34.38"	50	
25	03°18'30.30"	60°35'59.82"	25	
26	03°18'31.08"	60°35'58.98"	35	7th Knitting game
27	03°18'32.28"	60°35'57.24"	40	
28	03°18'29.94"	60°35'57.30"	50	
29	03°18'33.84"	60°35'55.62"	25	
30	03°18'36.12"	60°35'54.96"	35	8th Knitting game
31	03°18'37.98"	60°35'53.22"	40	
32	03°18'35.34"	60°35'54.78"	50	
33	03°18'14.76"	60°35'58.86"	25	
34	03°18'23.28"	60°35'59.70"	35	9th Knitting game
35	03°18'13.98"	60°36'0.66"	40	
36	03°18'20.64"	60°35'59.52"	50	
37	03°12'44.40"	60°38'19.14"	25	
38	03°13'41.70"	60°38'39.30"	35	10th Knitting game
39	03°13'45.36"	60°38'35.46"	40	
40	03°13'43.92"	60°38'36.84"	50	
41	03°13'43.14"	60°38'24.30"	25	
42	03°13'46.92"	60°38'34.32"	35	11 ° Mallet game
43	03°13'38.52"	60°38'27.30"	40	
44	03°13'28.56"	60°38'26.04"	50	
45	03°14'4.98"	60°38'99.68"	25	
46	03°13'58.44"	60°38'26.52"	35	12th Knitting game
47	03°13'58.56"	60°38'29.16"	40	
48	03°14'3.18"	60°38'20.76"	50	

Table 5.2: Ichthyofauna sampling sites in the Miriti APA.

The number of sample sites in the Miriti APA area was previously planned and defined, as the research horizon had to be delimited and therefore in such a way that we could obtain the expected results in the time we had to carry out the research.

According to VIALI (2016), the process of choosing a population involves risks, since decisions are made about the whole population on the basis of just one part of it. "Probability theory" tells us that it can be used to provide an idea of the risk involved, i.e. the error made by using a sample instead of the whole population, provided, of course, that the sample is selected using probabilistic criteria.

5.3. Effort and sampling efficiency (active search for animals)

The methodologies adopted for the active search for the animals were based on the two IPAAM Terms of Reference used to support technical and scientific studies for the inventory of fauna and the preparation of reports on the results of fauna surveys for the purposes of environmental licensing in Amazonas, as shown in Annexes 2 and 3, which are based on Normative Instruction No. 146 of January 11, 2007 of the Brazilian Institute of the Environment and Renewable Natural Resources - IBAMA.

After opening the transects, the animals were identified in a variety of places. Reptiles and amphibians were found in environments or hiding places such as: fallen trunks, rocks, tree bark, bushes, stones, holes or depressions in the ground, the banks of small streams, rivers and streams.

In the case of mammals, the search for specimens of fauna took place around areas where there were traces of animals drinking water, where there were trees that were fruiting, places of possible shelter or hiding places for predators, such as burrows, hollows of fallen trees, holes in the ground or in tree trunks.

Birds have also been found in trees during fruiting periods, in nests made in the upper part of the stem (hole), such as the woodpecker, the bacural and others, as well as observed in their nests in the branches or between the leaves as a mimicry strategy (camouflage). But we also found individuals under or in thickets of grass, maliça and jurubeba, using strategies to protect themselves from predators, as is the case with the common turtledove, juriti, galega and jaçanâ.

The fish were catalogued in various areas, such as around the aquatic macrophytes, under and around the flooded forest, near the riverbank, in areas where the rapids were intense, as well as in places where there was a lot of calm.

5.3.1. Direct and indirect identification of fauna

For the field research to catalog the primary fauna data, we looked for indications (traces) that an animal had been on the site or even passed through the area, either on the day of cataloging or on other nearby days, leaving its marks, impressions and characteristics on the site.

The data cataloged was both indirect and direct. Some animals were cataloged only indirectly, others directly and still others through two means of identification, indirect and direct, simultaneously.

5.3.1.1. Identification through indirect data

The means by which the animals were found through indirect data in the area of the sampling sites were: feces, feathers, gnawed fruit, carcasses, burrows (in the ground, in the trunk or stem of trees), active and abandoned nests, fish eggs in the water, bird eggs, tadpoles, vocalizations, footprints (figure 5.7).

Figure 5.7: Identification of fauna through indirect data. **a)** Woodpecker hole in the stem of a tree in a flooded area - Field cataloguing. **b)** Feathers found under vegetation. **c)** Fruit eaten by a rodent. **d)** Mammal's head. **e)** Bird feather found on a yurt in the forest. **f)** Mammal's burrow. **g)** Bird's nest in a flooded area. **h)** Fish eggs in an area with few rapids. **Source:** Technical field collection - Solange B. Damasceno (2013-2015).

5.3.1.2. Identification through direct data

The direct fauna data was catalogued by directly visualizing the animals and identifying them according to their respective vertebrate groups (avifauna, herpetofauna, ichthyofauna and mastofauna (figure 5.8).

As the animals were catalogued in the environments, they were also quantified, observing their characteristics, their characteristic and/or usual locations, as well as the times at which they were seen.

a
b
c
d
e
f
g

Figure 5.8: Identification of fauna using direct data. **a)** Cuiu-cuiu caught in a mesh upstream of the Miriti resort. **b)** Dogfish caught in a mesh downstream of the Miriti resort. **c)** Lizard on macrophyte near a house. **d)** Iguana on the banks of the river. **e)** Canuaru toad on a plateau. **f)** Green ray caught under macrophytes. **g)** Great egret in search of food. **h)** Sloth high up in the flooded forest. **Source:** Technical field collection - Solange B. Damasceno (2013-2015).

5.3.1.3. Identification through indirect and direct data

During the field research, it was possible to identify the animals both indirectly and directly (figure 5.9), i.e. by observing a trace of the animal, as well as by seeing the animal itself.

Figure 5.9: Identification of fauna through indirect and direct data. **a)** Japiim and nests. **b)** Bacural and nest. **c)** Socói catching food. **d)** Unidentified Piabas. **e)** Egg found on the ground in a slope area. **f)** Hummingbird in the nest during the hatching period. **Source:** Technical field collection - Solange B. Damasceno (2013-2015).

CHAPTER 6: RESULTS

In the Miriti Environmental Protection Area, the methodologies applied to catalog the existing fauna for the development of this research showed that the diversity of individuals corresponds to that of "terra-firme forests and Amazonian plains" found in other similar environments, Therefore, the exotic, endemic, vulnerable and rare species listed in the two Ordinances published by the Ministry of the Environment/Chico Mendes Institute for Biodiversity Conservation (MMA/ICMBIO) were taken into account: MMA Ordinance No. 444, of December 17, 2014 (terrestrial species and aquatic mammals): 698 taxa; and MMA Ordinance No. 445, of December 17, 2014 (fish and aquatic invertebrates): 475 taxa (ICMBIO, 2016).

The diversity of fauna catalogued in the Miriti APA amounted to 163 species from 80 different families (figure 6.1).

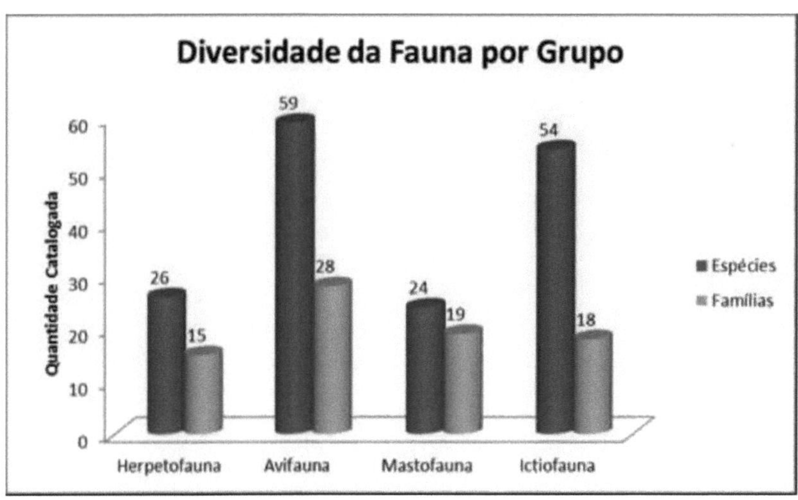

Figure 6.1: Diversity of Catalogued Fauna.

Of the 163 species of fauna identified in the Miriti APA, 26 were from the Herpetofauna (16%), 59 from the Avifauna (36%), 24 from the Mastofauna (15%) and 54 from the Ichthyofauna (33%). And of the 80 fauna families identified in the survey, 15 were Herpetofauna (19%), 28 Avifauna (35%), 19 Mastofauna (24%) and 18 Ichthyofauna (22%)

(table 6.1).

Groups	Species	% of Species by Group	Families	% of families by group
Herpetofauna	26	16%	15	19%
Avifauna	59	36%	28	35%
Mastofauna	24	15%	19	24%
Ichthyofauna	54	33%	18	22%
Total	163	100	80	100

Table 6.1: Fauna Identified by Species and Family.

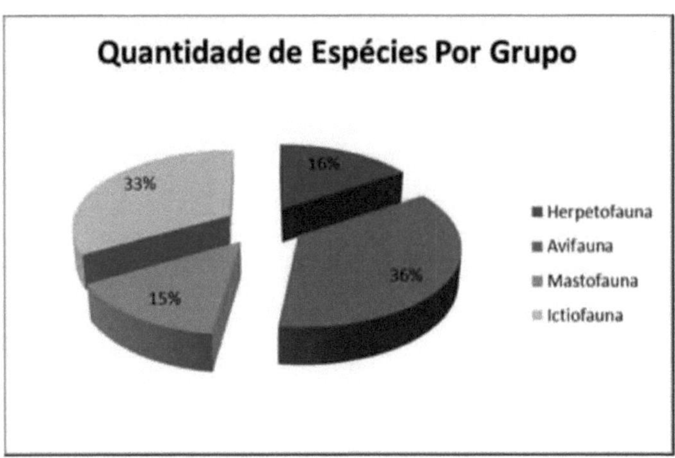

Figure 6.2: Fauna Species Identified by Group in %.

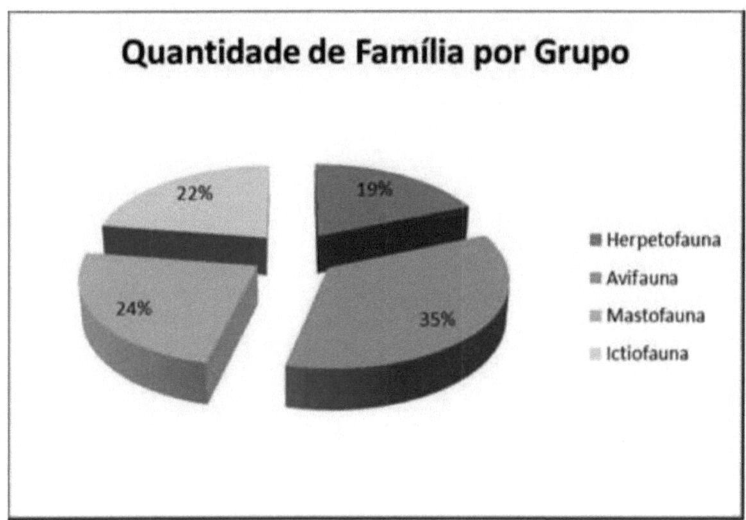

Figure 6.3: Fauna Family Identified by Group in %.

6.1. Identification of fauna by group and by individual

Tables 6.2, 6.3, 6.4 and 6.5 list the diversity of the fauna identified in the survey in the Miriti APA, specimen by specimen in their respective fauna groups, herpetofauna, avifauna, Mastofauna and Ichthyofauna, respectively, identifying the families, species and popular name of each individual.

Item	Family	Species	Popular Name
01	Hylidae	*Hyla minuta*	Yellow ra
02	Hylidae	*Hyla lanciformis*	Ra brown
03	Hylidae	*Hyla granosa*	Green Ra
04	Bufonidae	*Bufo proboscideus*	Toad leaf
05	Bufonidae	*Bufo marinus*	Cururu frog
06	Leptodactylidae	*Leptodactylus pentadactylus*	Canuaru frog
07	Alligatoridae	*Caiman crocodilus*	Alligator Tinga
08	Alligatoridae	*Paleosuchus trigonatus*	Tiritiri alligator
09	*Alligatoridae*	*Melanosuchus niger*	Alligator
10	Boidae	*Boa constrictor*	boa constrictor
11	Boidae	*Epicrates cenchria*	Rainbow boa
Item	**Family**	**Species**	**Popular Name**
12	Boidae	*Eunectes murinus*	Sucuri
13	Boidae	*Corallus caninus*	Parrot snake
14	Colubridae	*Pseustes sulphureus*	Egg-eating snake
15	Colubridae	*Chironius exoletus*	Cobra cipó
16	Chelidae	Rhinoclemmys punctularia	Perema

17	Chelidae	*Chelus fimbriata*	Matâmatâ
18	Elapidae	Micrurus corallinus	Coral snake
19	Iguanidae	*Iguana iguana*	Iguana
20	Podocnemidae	*Podocnemis sextuberculata*	Iace
21	Pelomedusidae	*Podocnemis expansa*	Turtle
22	Pelomedusidae	*Podocnemis unifilis*	Tracajâ
23	Scincidae	Mabuya nigropunctata	Lizard
24	Testudinidae	*Geochelone denticulata*	Jabuti
25	Teiidae	*Tupinambis teguixin*	Jacuraru
26	Teiidae	*Ameiva ameiva*	Sweet beak or green calango
27	Viperidae	Bothrops atrox	Jararaca
28	Viperidae	*Lachesis muta*	Surucucu snake

Table 6.2: Herpetofauna (Amphibians).

Item	Family	Species	Popular name
01	Accipritidae	*Buteo magnirostris*	Caryngeal warbler
02	Accipitridae	*Busarellus nigricollis*	Red jay
03	Accipitridae	*Rostrhamus sociabilis*	Caraway
04	Accipitridae	*Harpy harpy*	Royal caribou
05	Apodidae	*Panyptila cayennensis*	Swallow
06	Anatidae	*Cairina moschata*	Wood duck
07	Ardeidae	*Tigrisoma lineatum*	Socó-boi
08	Ardeidae	*Egretta thula*	Little egret
09	Ardeidae	*Cosmerodius albus*	Great egret
10	Ardeidae	*Butorides striatus*	Punch
11	Bucconidae	*Chelidoptera tenebrosa*	Beak
12	Caprimulgidae	*Chordeiles leucopyga*	Bacurau
13	Cerylidae	*Ceryle torquata*	Ariramba grande
14	Ciconiidae	*Coragyps atratus*	Common black vulture
Item	Family	Species	Popular name
15	Ciconiidae	*Cathartes aura*	Red-headed vulture
16	Columbidae	*Columba cayennensis*	Galegao
17	Columbidae	*Leptotila rufaxilla*	Juriti
18	Columbidae	*Columbina passerina*	Galician
19	Columbidae	*Columbina talpacoti*	Common turtledove
20	Cracidae	*Penelope superciliaris*	Jacu
21	Crotophagidae	*Crotophaga major*	Anu-coroca
22	Crotophagidae	*Crotophaga ani*	Black Anu or Anum
23	Cuculidae	*Piaya cayana*	Ticoa
24	Dendrocygnidaand	*Dendrocygna autumnalis*	Marreca
25	Eurypygidae	*Eurypyga helias*	Pavaozinho
26	Fringillidae	*Scaphidura oryzivora*	Graùna
27	Fringillidae	*Sryzoborus angolensis*	Curió
28	Fringillidae	*Thraupis episcopus*	Crab

Item	Family	Species	Popular name
29	Fringillidae	*Psarocolius decumanus*	Japu
30	Fringillidae	*Sicalis flaveola*	Canary
31	Fringillidae	*Ramphocelus carbo*	Red kite
32	Fringillidae	*Tachyphonus rufiventer*	Pipira
33	Fringillidae	*Cacicus cela*	Japanese bearded man
34	Fringillidae	*Paroaria gularis*	Galo da Campina
35	Jacanidae	*Jacanajaçana*	Jaçana / Piaçoca
36	Opisthocomida and	*Opisthocomus hoazin*	Gypsy
37	Phalacrocoracidae	*Phaslacrocorax brasilianus*	Diving
38	Picidae	*Celeus elegans jumana*	Red woodpecker
39	Picidae	*Campephilus melanoleucos*	Red-cockaded woodpecker
40	Psittacidae	*Festive amazon*	Parrot
41	Psittacidae	*Ara manilata*	Maracana
42	Psittacidae	*Ara macao*	Macaw Canga
43	Psittacidae	*Ara ararauna*	Canindé macaw
44	Psittacidae	*Forpus Xanthopterygius*	Curica
45	Ramphastidae	Ramphastos vitellinus	Yellow-breasted toucan
46	Ramphastidae	*Ramphastos tucanus cuvieri*	White-breasted toucan
47	Rallidae	Porphyrio martinicus	Water chicken - blue
48	Rallidae	*Aramus guaraùna*	Carao
Item	**Family**	**Species**	**Popular name**
49	Scolopacidae	*Solitary trinket*	Solitaire torch
50	Scolopacidae	*Carlidris fuscicollis*	Torch
51	Tinamidae	*Tinamus guttatus*	Inhambu- chicken
52	Tyrannidae	*Empidonomus varius*	Peitica
53	Tyrannidae	*Pitangus sulphuratus*	Birds of a feather
54	Tyrannidae	*Myiophobus fasciatus*	Filipe
55	Threskiornithidae	*Mesembrinibis cayennensis*	Corocoró
56	Trogonidae	*Trogon curucui*	Blue-crowned Surucuà
57	Trochilidae	*Chlorostilbon mellisugus*	Green hummingbird

Table 6.3: Avifauna (Birds).

Item	Family	Species	Popular name
01	Atelidae	*Alouatta macconnelli*	Guariba monkey
02	Bradypodidae	*Bradypus tridactylus*	Sloth
03	Cebidae	*Saimiri sciureus*	Smelling Monkey
04	Cebidae	*Cebus apella*	Nail Monkey
05	Cebidae	*Cebus albifrons*	Cairara monkey
06	Caviidae	*Hydrochoerus hydrochoeris*	Capybara
07	Cervidae	*Mazama americana*	Fallow deer
08	Cricetidae	*Oligoryzomys microtis*	Bush mouse

Item	Family	Species	Popular name
09	Cuniculidae	*Cuniculus paca*	Paca
10	Dasyproctidae	*Dasyprocta fuliginosa*	Cutia
11	Dasypodidae	*Priodontus maximus*	Canasta armadillo
12	Dasypodidae	*Dasypus novemcinctus*	Chicken armadillo
13	Delphinidae	*Sotalia fluviatilis*	Tucuxi dolphin
14	Didelphity	*Didelphis sp*	Mucura or gambà
15	Erethizontidae	*Coendou prehensilis*	Coandu
16	Felidae	*Leopardus pardalis*	Maracajà cat
17	*Felidae*	*Puma concolor*	Grizzly jaguar
18	Felidae	*Panthera jaguar*	Jaguar
19	Iniidae	*Inia geoffrensis*	Pink dolphin
20	Megalonychidae	*Choloepus didactylus*	Real laziness
21	Mustelidae	*Eira barbara*	Irara
22	Mustelidae	*Otter longicaudis*	Otter
23	Mustelidae	*Pteronura brasiliensis*	Ariranha
24	Phyllostomidae	*Lonchophylla thomasi*	Bat
25	Tayassuidae	*Tayassu pecari*	Bison pig
Item	**Family**	**Species**	**Popular name**
26	Tapiridae	*Tapirus terrestris*	Tapir
27	Tayassuidae	*Pecari tajacu*	Pork and fish
28	Pitheciidae	*Callicebus cupreus*	Zogue monkey
29	Procynidae	*Nasua nasua*	Quati

Table 6.4: Mastofauna (Mammals).

Item	Family	Species	Popular Name
01	Anostomidae	*Leporinus friderici*	Fat-headed arachoo
02	Anostomidae	*Schizodon fasciatus*	Common arachnid
03	Anostomidae	*Laemolyta varia*	Penguin arachnid
04	Anostomidae	*Rhytiodus microlepis*	Black-bellied woodpecker
05	Arapaimidae	*Arapaima gigas*	Pirarucu
06	Callichthyidae	*Callichthys callichthys*	Tamautâ
07	Characidae	*Brycon melanopterus*	Jatuarana
08	Characidae	*Brycon amazonicus*	Matrinxa
09	Characidae	*Mylossoma duriventre*	Butterfish
10	Characidae	*Serrassalmus spilopleura.*	Yellow piranha
11	Characidae	*Pygocentrus nattereri*	Cashew piranha
12	Characidae	*Serrasalmus gouldingi*	White piranha
13	Characidae	*Pristobrycon striolatus*	Piranha pacu
14	Characidae	*Serrasalmus rhombeus*	Black piranha
15	Characidae	*Triportheus elongatus*	Long sardine
16	Characidae	*Colossoma macropomum*	Tambaqui
17	Characidae	*Characidium fasciatum*	Cigars
18	Cichlidae	*Pterophyllum scalare*	Cararâ flag

Item	Family	Species	Popular Name
19	Cichlidae	*Astronotus crassipinnis*	Acarâ açu
20	Cichlidae	*Chaetobranchopsis orbicularis*	White sugar

Item	Family	Species	Popular Name
21	Cichlidae	*Satanoperca jurupari*	Acarà papa terra (Acarâ-jarupari)
22	Cichlidae	*Uaru amphiacanthoides*	Acara bararuà
23	Cichlidae	*Hypselecara temporalis*	Acara cascudo (Purple Acarà)
24	Cichlidae	*Crenicichla cincta*	Jacundà
25	Cichlidae	*Cichla SP*	Tucunaré
26	Cichlidae	*Aequidens pallidus*	Acarà beré-beré
27	Cichlidae	*Mesonauta festivus*	Acarà boari
28	Cynodontidae	*Cynodon gibbus*	Dogfish
29	Cynodontidae	*Hydrolycus scomberoides*	Pirandirà
30	Curimatidae	*Potamorhina latior*	Common white
31	Curimatidae	*Psectrogaster amazonica*	Chest of steel (white skinny girl)
32	Clupeidae	*Pellona castelnaeana*	Yellow apapa
33	Doradidae	*Acanthodoras spinosissimus*	Reco-reco
34	Doradidae	*Oxydoras niger*	Cuiù-cuiù
35	Erythrinidae	*Hoplerythrinus unitaeniatus*	Jiju
36	Erythrynidae	*Hoplias malabaricus*	Moth
37	Gymnotidae	*Electrophoridae electrictus*	Poraquê
38	Hemiodontidae	*Argonectes longiceps*	Orana
39	Lepidosirenidae	*Lepidosiren paradoxa*	Pirambóia
40	Loricariidae	*Platydoras costatus*	Bacu
41	Loricariidae	*Liposarcus pardalis*	Bodó
42	Loricariidae	*Loricaria fc.simillima*	Bodó pipe
43	Loricariidae	*Platydoras costatus*	Rebeca

Item	Family	Species	Popular Name
44	Osteoglossidae	*Osteoglossum bicirrhosum*	Aruana
45	Pimelodidae	*Hypophytalmus edentatus*	Maparà
46	Pimelodidae	*Pseudoplatystoma fasciatum*	Surubim

47	Pimelodidae	*Leiarius marmoratus*	Jandià
48	Pimelodidae	*Pimelodella cristata*	Mandi
49	Pimelodidae	*Pseudoplatystoma tigrinum*	Caparari
50	Pimelodidae	*Sorubim lima*	Duckbill
51	Prochilodontidae	*Prochilodus nigricans*	Curimata
52	Prochilodontidae	*Semaprochilodus taeniurus*	Jaraqui-thin scale
53	Prochilodontidae	*Semaprochilodus insignis*	Jaraqui-thick-scaled
54	Sciaenidae	*Plagioscion squamosissimus*	Hake

Table 6.5: Ichthyofauna (Fish).

6.2. Alpha diversity

According to the Shannon-Wiener index in which H'= - $\sum$ Pi Log pi), calculated from the proportion of species in relation to the total number of individuals catalogued in the research into the diversity of fauna in the Miriti APA, there was a low to medium diversity of species at the site in relation to the diversity of Amazonian fauna. This index has an advantage over the Margalef, Gleason and Menhinick indices, as it is suitable for random samples of species from a community or sub-community of interest (Rodrigues, Costa William. Riqueza e Diversidade de Espécies, USS, 2012).

6.3. Beta diversity

According to the total diversity index sv (i), the following equation was used: TD= $\sum$Wi [Pi (1-Pi)]. Where: wi is the weight given to the function, which expresses the importance given to species i in the overall classification of regional diversity; pi is the relative frequency. Wi is calculated using the expression: wi=1/pi. The faunal diversity catalogued in the research can be estimated as a variation of species, where it was considered to be medium to low according to other areas, but the result obtained could not be compared with others, as no studies were found in the literature on other Environmental Protection Areas in the region, analyzing beta diversity with similarity.

6.4. Relevant exotic fauna species

The fauna inventoried in the Miriti APA, described in Table 2, were not found to be exotic or endemic, but we consider them to be highly relevant, given that the community uses them for subsistence.

6.5. Endangered species

According to the current Lists of Brazilian Fauna Species Threatened with Extinction (MMA Ordinances No. 444/2014 and No. 445/2014), there are 1,173 species.

Of the total of 163 different species of fauna cataloged, twelve (12) species are on the Official List of Endangered Brazilian Fauna (MMA Ordinances No. 444/2014 and No. 445/2014), *Saimiri sciureus* (scent monkey), *Priodontes maximus* (canastra armadillo), *Leopardus pardalis* (maracajà cat), *Puma concolor* (brown jaguar), *Panthera onca* (jaguar), *Pteronura brasiliensis* (giant otter), *Crenicichla cincta* (jacundà), surubim, *Semaprochilodus insignis* (thick-bladed *tamarin*), *Semaprochilodus taeniurus* (thin-bladed tamarin), *Colossoma macropomum* (tambaqui) and *Arapaima gigas* (pirarucu).

CHAPTER 7: DISCUSSIONS

The municipality of Manacapuru is part of the metropolitan region of Manaus, which includes the municipalities of Novo Airao, Iranduba, Manaus, Presidente Figueiredo, Rio Preto da Eva, Itacoatiara and Careiro da Vàrzea. Its mission is to carry out public functions which, by their nature, require cooperation between these municipalities to solve common problems, This legitimizes its existence in political-institutional terms, as well as allowing the public authorities to act in a more integrated way to meet the needs of its residents, who are identified with the institutionalized territorial area (figure 7.1).

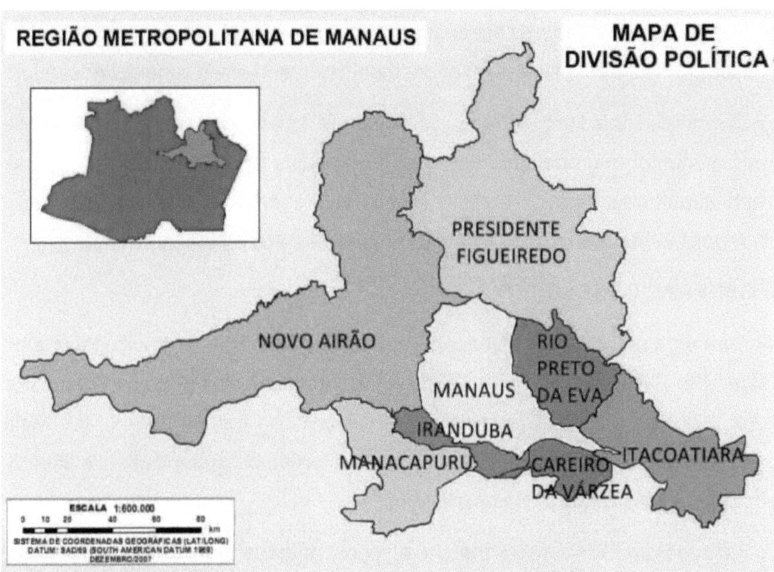

Figure 7.1: Map of the Political Division of the Metropolitan Region of Manaus. **Source**:

IBGE - Demographic Census - 2010.

The condition of the municipality of Manacapuru as a participant in a metropolitan region is guaranteed by the Federal Constitution of 1988, in its Article 25, paragraph 3 and recognized by the Court of Justice of the State of Amazonas, in the Electronic Journal of Justice, year III, edition 624, on November 8, 2010.

"The grouping of the municipality of Manacapuru with the other neighbouring municipalities was due to the objective of integrating the organization, planning and execution of public functions of common interest."

This responsibility and attributions, which have been given to them as a result of their role as a metropolitan area, must necessarily be put into practice.

And since the implementation of structural projects such as the Coari Manaus gas pipeline, the bridge over the Rio Negro and currently the duplication of the AM 070 highway, Manacapuru has had to undergo restructuring in various aspects and sectors, including municipal environmental management.

Among the many agendas that need to be rethought and restructured, we would highlight the environmental agenda, the body of which is the Municipal Secretariat for the Environment and Tourism - SEMANTUR - so that it can effectively fulfill its role as the body that manages and implements municipal environmental policy.

The Miriti APA is a reflection of the neglect and absence of command and control by public authorities, in discussions, supervision, regulation, monitoring and environmental education.

The biological diversity in the Miriti APA is in fact surviving and coexisting between the natural and environmental degradation. However, the impacts are increasing day by day and serious and structuring decision-making measures need to be taken in this area, especially with regard to the management of water resources and all that surrounds it.

CHAPTER 8: GENERAL CONCLUSIONS

As it is a peri-urban area of the city of Manacapuru, the Miriti APA is in a transitional phase. On the right side of the Miriti basin, a large part of it is urbanized and/or in the process of being urbanized, including the headwater areas. In the area on the left bank of the Miriti basin, we can still find areas with fairly preserved riparian forest, but urbanization is already part of it, specifically in the vicinity of Balneàrio Miriti.

We can also conclude from the fieldwork that the diversity of fauna found there has uniform characteristics throughout. With regard to certain groups of fauna (avifauna, mastofauna and herpetofauna), we can say that there is a constancy in their identification throughout the route surveyed, but with regard to the ichthyofauna, it was observed that in the region upstream of the Miriti bathing resort there is a diversification of the species catalogued in this research, which are therefore bottom-dwelling (leather) fish, called catfish.

This is probably because of the type of habitat in the area, such as flooded forest, because it is a perennial area with a lacustrine character, calm waters and because it is close to the headwaters. These types of fish usually migrate upstream to spawn during the drought or the beginning of the rains.

The downstream part of Balneàrio Miriti, being a well-urbanized region, has a different character because it is more dynamic due to various factors, including: the flow of water from upstream towards the downstream, the various waters coming from the branches of the Miriti River and the Calado lakes and the flow of water from the Solimoes River, due to the seasonal nature of the Amazon.

In the areas that border the Miriti River and meet the Solimoes River during the state's seasonal flooding, the Solimoes advances over the Miriti and floods the flat or lowland areas (figure 8.1; a, b, c, d), causing great damage and discomfort to the community living in the outlying areas. However, this phenomenon of flooding and drought is extremely important for life in the Miriti basin in the context of gene flow.

Figure 8.1: Flow of the Solimoes River under the Miriti River during the flood. **a** and **b** Homes flooded during the flood. **c** and **d** Advance of the Solimoes River over the Miriti River. **Source**: Technical field collection - Solange B.

Damasceno (2013-2015).

We also found that during the flood period, when the waters of the Solimoes River enter the Miriti basin, when the piracema phenomenon occurs ("*migratory movement of fish in the direction of river sources, for the purpose of reproduction*"), there is indiscriminate fishing with large, non-selective nets (figure 8.2; a, b), probably causing direct or indirect impacts on the reproduction of the local ichthyofauna and species diversity. What should be regulated in the area is apparently not controlled at all by the command and control bodies.

Figure 8.2: Fishing during the Spawning Season on the Miriti River. **a** and **b** - Artisanal and commercial fishing during the flood season on the Miriti River. **Source**: Technical field collection - Solange B. Damasceno (2013-2015).

We also found that in this area of the Miriti basin are located the two surface water collection points that supply the city of Manacapuru. One of the surface water collection points is located in the São Francisco neighborhood and the second in the Liberdade neighborhood, both managed by SAAE (Serviço Autónomo de Agua e Esgoto) (figure 8.3).

It is worth noting that during the flood period the Miriti River is robust and has an abundance of water, but during the dry season the quality of the water is greatly compromised, not least because of the organic discharge that is deposited directly into the river or in its lowland areas.

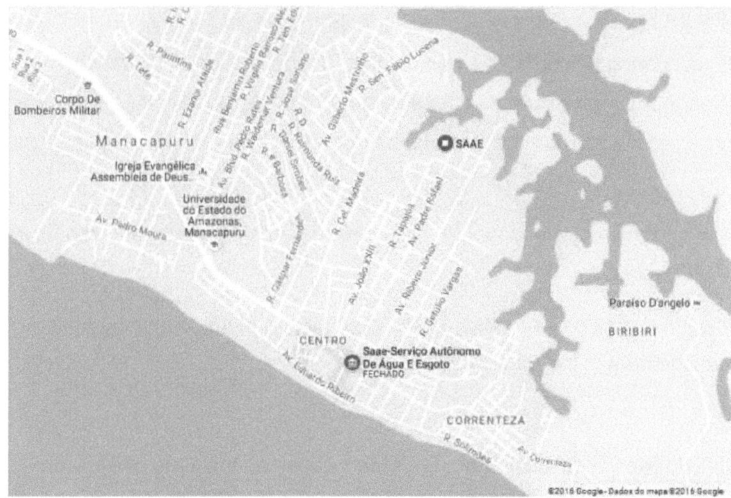

Finally, one of the biggest impacts on wildlife diversity in the APA area was observed in the Miriti river channel in the area that connects the Calado lakes. This area is obstructed by a bridge built of clay material linking one point of land to another on a large private property, damming up an area and preventing species from circulating.

CHAPTER 9: RECOMMENDATIONS

Protection of the characteristic igapó areas, both upstream and downstream of Balneàrio Miriti, as the population uses them for artisanal fishing, both for survival and for selling fish.

Planning of sports and leisure areas, including ecological tourism (fishing, alligator spotting, canoeing and others).

Control of indiscriminate fishing with non-selective nets.

Make a fishing agreement between residents and stakeholders so that the species on the closed list don't become extinct.

Deploying in the area, apparently there is no control by the command and control bodies.

Re-evaluation of the surface water collection points that supply the city of Manacapuru, as the Miriti River has gradually silted up over the years due to the environmental impacts of the urban area's disorderly growth under the riparian forest areas, thus compromising the quality of the surface water.

Insert water quality monitoring points for the well-being of the population of the municipality of Manacapuru, given the compromised health caused by water-borne diseases and other pollutants that may be percolating into the food chain through living organisms that serve as human food.

Implementation of environmental monitoring in the area to control impacts and preserve biological diversity in the area.

Creation of the Miriti River Basin Committee as the regulating and implementing body for water resources management in the Miriti basin.

BIBLIOGRAPHIES

APRF.Tourism. Miriti do Rio waterfront in the city of Manacapuru in the state of Amazonas, in the northern region of Brazil. Retrieved from http://mochileiro.tur.br/manacapuru.htm.

Revised on: 17/10/15.

Barreto, M. W. A. (2009). *Economic and Socio-environmental Aspects Relative to the Sustainability of Fisheries Resources in the Piranha Sustainable Development Reserve (RDS)* - Manacapuru-AM. Federal University of Amazonas. Manaus.

Bernarde, P. S. (2012). *Amphibians and reptiles: Introduction to the Study of the Brazilian Herpetofauna*. Curitiba: Anolisbooks.

Bernardino, F. R. and Omena, J. R. S. (2002). *Aves da Amazônia - Guia do Observador*. Editora Paper.

Bini, E. (2009). *Birds of Brazil - Practical Guide*. Lages. 1ª Edition. p. 462.

Capobianco, J. P. R.; Verissimo, A.; Moreira, A.; Sawyer, D.; Santos, I. and Pinto, L. P. (2001). *Biodiversity in the Brazilian Amazon: Assessment and Priority Actions for Conservation, Sustainable Use and Benefit Sharing*. Liberdade Station. Socio-Environmental Institute, Sao Paulo, SP.

Forsberg, B. R.; Lima, A.; Martinelli, L. A. and Victoria, B. R. L. (1993). Autotrophic carbon sources for fish of the central Amazon. *Ecology, 74*, 643- 652.

Goeldi, E.   (1981). *Album de Aves Amazónicas*. 2nd ed. Brasilia. Editora Universidade de Brasilia, CNPq.

Government of the State of Rio Grande do Sul (2008). *RS Biodiversity Project*. Planning and Management Secretariat. Project for the Conservation of Biodiversity as a Contributing Factor to the Development of the State of Rio Grande do Sul, RS.

Amazonas State Government, Manacapuru City Hall - PMM (2012). *Project for the Implementation of the Manacapuru Conservation Unit Program*. Municipal Secretariat for the Environment, Sustainable Development and Tourism. Biological Report of the Miriti Environmental Protection Area, Manacapuru, Am.

Government of the State of Amazonas (2016), *Termo de Referência para Elaboração de Inventàrio e Levantamento de Fauna*, IPAAM.

Government of the State of Amazonas (2016), *Termo de Referência para Elaboração do Relatório dos Resultados do Levantamento de Fauna*, IPAAM.

Brazilian Institute of Geography and Statistics - IBGE (2016). *Amazonas, Manacapuru, Municipal Human Development Index - IDHM*. Atlas Brazil 2013 United Nations Development Program - 1991, 2000 and 2010. Retrieved from http://www.cidades.ibge.gov.br/xtras/temas.php?lang=&codmun=130250&id tema=118&search=amazonas|manacapuru|%C3%8Dndice-de- desenvolvimento-humano-municipal-idhm-.

Ministry of Planning, Budget and Management, Brazilian Institute of Geography and Statistics - IBGE (2010). *2010 Demographic Census.*

Subnormal agglomerations, Territorial information. Rio de Janeiro, p. 1-251.

Brazilian Institute for the Environment and Renewable Natural Resources - IBAMA (2007), *Normative Instruction No. 146.*

Chico Mendes Institute for Biodiversity Conservation - ICMBIO/MMA (2009). *ICMBIO Legislation Series*, Volume 1. Law No. 9.985, of July 18, 2000. National System of Nature Conservation Units - SNUC.

Junk, W. J. and Silva, C. J. (1997). *Mammals, reptiles and amphibians. The Central Amazon Floodplain, Ecology of Pulsing System.* Springer Verlag, Berlin, p. 409-417.

Lima, A. P.; Keller C. and Magnussom, W. E. (2005). *Guide to the Frogs of the Adolpho Ducke Reserve.* Manaus: Attema Design Editorial.

Ministry of the Environment, Chico Mendes Institute for Biodiversity Conservation - ICMBIO (2016). *Endangered Fauna Ordinances. Ordinance n^0 444 of December 17, 2014.* Retrieved from http://www.icmbio.gov.br/portal/faunabrasileira?id=6706:portarias-fauna- threatened

Ministry of the Environment, Chico Mendes Institute for Biodiversity Conservation - ICMBIO (2016). *Endangered Fauna Ordinances. Ordinance n^0 445. Of December 17, 2014.* Retrieved from http://www.icmbio.gov.br/portal/faunabrasileira?id=6706:portarias-fauna-threatened

Mafham, K. P.; Marven, N. and Harvey, R. (1998). *Insects, Spiders and Snakes.* Ediçâo 70. Lisbon, Portugal.

Ministry of the Environment - MMA (2002). *Assessment and identification of priority areas and actions for the conservation, sustainable use and sharing of the benefits of biodiversity in Brazilian biomes.* Brasilia. SBF-404 p.

Oliveira, M. L.; Baccaro, F. B.; Neto, R. B. and Magnusson, W. E. (2008). *Ducke Reserve: Amazonian biodiversity through a grid.* National Institute for Amazonian Research - INPA. Manaus: Attema Design Editorial.

Panzu, A. N. S. (2015). *Institutional trajectory through its scientific practices, 1954-1975.* Master's dissertation in History. National Institute for Amazonian Research - INPA. Manaus,

Am: UFAM.

Planalto - Presidency of the Republic (2016). *News: Brazil intensifies actions to protect fauna and flora at risk of extinction.* Retrieved from http://www2.planalto.gov.br/noticias/2015/06/brasil-intensifica-acoes-para- protect-fauna-and-flora-at-risk-of-extinction

National Program for the Development of Amateur Fishing - PNDPA (1999). Brazilian Sportfishing *Guide.* Brasilia.

Reis, N. R.; Peracchi, A. L.; Fregonezi, M. N. and Rassaneis, B. K. (2010). *Mammals of Brazil - Identification Guide.* Rio de Janeiro. Publisher: Technical Books.

Rodrigues, C. W. (2012). *Species Richness and Diversity,* USS.

Santos, G. M.; Ferreira, E. J. G. and Zuanon, J. A. S. (2006). *Commercial Fish of Manaus.* IBAMA - AM, ProVàzea.

Vitt, L.; Magnusson, W. E.; Pires, T. C. A. and Lima, A. P. (2008). *Lizard Guide to the Adolpho Ducke Reserve, Central Amazonia.* Manaus: Atlema Design Editorial.

ANNEXES

Annex 1 -TR - Term of Reference for Floristic Inventory for the purpose of obtaining the IPAAM Single Environmental License for Vegetation Suppression.

TERMO DE REFERÊNCIA
INVENTÁRIO FLORÍSTICO PARA FINS DE OBTENÇÃO DA
LICENÇA AMBIENTAL ÚNICA DE SUPRESSÃO DE VEGETAÇÃO

O Instituto de Proteção Ambiental do Amazonas – IPAAM, autarquia estadual criada pela Lei nº. 2.367 de 14.12.95 e instituída pelo Decreto nº. 17.033, de 11.03.96, com sede à Rua Mário Ypiranga Monteiro, nº. 3.280, Parque Dez de Novembro em Manaus – AM, detalha a forma de elaboração e apresentação do Inventário Florístico para fins de obtenção Licença Ambiental Única – LAU de Supressão de Vegetal.

Considerando a necessidade de disciplinar os procedimentos relativos às Licenças de Supressão Vegetal de **empreendimentos de interesse público ou social** submetidos ao licenciamento ambiental junto a este IPAAM, segue Termo de Referência.

1. INFORMAÇÕES GERAIS

1.1. Identificação do Proprietário.
- Nome
- RG e CPF/CNPJ
- Endereço para correspondência
- Telefone para contato

1.2. Identificação do Elaborador e Executor
- Nome
- Endereço
- Telefone para contato

2. INVENTÁRIO FLORÍSTICO

A seguir, estão descritas os itens e sub-itens que servirão de orientação técnica para a elaboração do Inventário Florístico, devendo o interessado realizar e apresentar:

1. O levantamento florístico das espécies arbóreas, arbustivas, palmeiras arborescentes, em todos os estratos da vegetação (arbustivo e arbóreo);

2. O levantamento florístico deverá conter informação qualitativa e quantitativa da vegetação, objeto de supressão, devendo ainda contemplar informações acerca da família botânica, nomes científico e comum, hábito, tipo de vegetação, estrato e, quando for o caso, estado fenológico e número de tombamento;

3. Fica estabelecido o DAP (Diâmetro à altura do Peito) mínimo de 15 cm, para procedimento de levantamento de campo;

4. A localização das unidades amostrais no levantamento florístico (20 metros x 125 metros), deverá conter as coordenadas geográficas dos vértices;

5. A amostragem deverá atender as premissas da experimental florestal;

6. Poderá ser utilizada a equação de volume (m^3) – ln $V = -7,335 + 2,121*$lnDAP ($R^2=0,95$ / $S_{yx} = 0,27$). Fonte: Projeto Bionte – Biomassa e Nutrientes. 1997, pg. 93;

7. Tabela de volume por espécie (nome científico) e por produto (tora, escoramento, estacas, varas ou postes) a ser explorado e volume total por produto;

OBSERVAÇÃO: Em caso de intervenção em Área de Preservação Permanente – APP deverá o inventário florestal representar a realidade local, no que se refere ao volume e espécies.

3. RESPONSÁVEL TÉCNICO

O inventário florístico deverá ser realizado por profissional habilitado com apresentação do registro no Conselho de Classe, registro IPAAM e Anotação de Responsabilidade Técnica, que por sua vez será o responsável técnico pelos resultados apresentados.

4. APRESENTAÇÃO

O levantamento florístico deverá ser apresentado em meio físico e digital, de forma objetiva, e adequada à compreensão, de acordo com as normas da ABNT, devendo atender ao conteúdo estabelecido neste TR, e com a respectiva assinatura do técnico, contendo o nome completo, número de registro no órgão de classe e o número do cadastro como prestador de serviços na área ambiental, expedido por este IPAAM.

TERMO DE REFERÊNCIA PARA ELABORAÇÃO DE INVENTÁRIO E LEVANTAMENTO DE FAUNA

O levantamento de Fauna é utilizado nos trabalhos de estudos ambientais de uma determinada área tendo como finalidade a verificação das espécies presentes no local para orientação de tomada de decisão.

Este Termo de referência apresenta o conteúdo mínimo a ser contemplado na elaboração de atividades de levantamento de fauna.

1. Lista de espécies da fauna descritas para a localidade ou região , baseada em dados secundários, inclusive com indicação de espécies constantes em listas oficiais de fauna ameaçada com distribuição potencial na área do empreendimento, independentemente do grupo animal a que pertencem. Na ausência desses dados para a região, deverão ser consideradas as espécies descritas para o ecossistema ou macro região;

2. Descrição detalhada da metodologia a ser utilizada no registro de dados primários, que deverá contemplar os grupos de importância para a saúde pública regional, cada uma das classes de vertebrados, e classes de invertebrados pertinentes e incluindo informação quanto a periodicidade de revisão das armadilhas. Em caso de ocorrência no local do empreendimento, de focos epidemiológicos, fauna potencialmente invasora, inclusive doméstica ou outras espécies oficialmente reconhecidas como ameaçadas de extinção, o IPAAM poderá ampliar as exigências de forma a contemplá-las. A metodologia deverá incluir o esforço amostral para cada grupo em cada fitofisionomia, contemplando a sazonalidade para cada área amostrada;

3. Informação referente ao destino pretendido para o material biológico a ser coletado, com anuência da instituição onde o material será depositado, válida por até dois anos;

4. Mapas imagens de satélite ou foto da área, inclusive com avaliação batimétrica e altimétrica, contemplando a área afetada pelo empreendimento com indicação das fitofisionomias, localização e tamanho das áreas a serem amostradas;

5. Identificação da bacia e microbacias hidrográficas e área afetada pelo empreendimento. Deverão ser apresentados mapas com a localização do empreendimento e vias de acesso pré-existentes;

6. Descrição dos sítios amostrais com coordenadas geográficas ou planas (UTM);

7. Levantamentos de ictiofauna e invertebrados Aquáticos descritos para curso d'água e seus afluentes, baseada em dados secundários, indicando as espécies nativas, exóticas, reofílicas, de importância comercial, ameaçadas de extinção, sobreexplotadas, ameaçadas de sobreexplotação, endêmicas e raras. Na ausência de bibliografia específica, deverão ser consideradas as espécies descritas para a região hidrográfica;

8. Descrição detalhada da metodologia a ser utilizada para inventários de peixes, ictioplâncton, fitoplâncton, invertebrados aquáticos (zooplâncton e grandes grupos zoobentos), além dos bioindicadores de saúde pública e qualidade ambiental. As amostragens devem contemplar pelo menos a área de influência direta do empreendimento e a microbacia relacionada.

TERMO DE REFERÊNCIA PARA ELABORAÇÃO DO RELATÓRIO DOS RESULTADOS DO LEVANTAMENTO DE FAUNA

O Relatório é um documento que apresenta os resultados dos estudos técnicos e científicos do Levantamento de Fauna realizado. Este deverá contemplar ações e procedimentos que possibilitem o acompanhamento e a avaliação de suas atividades, como forma de garantir o efetivo alcance dos seus objetivos e metas, a partir da adoção dos métodos anteriormente preconizados.

Este Termo de referência apresenta o conteúdo mínimo a ser contemplado na elaboração do relatório dos resultados do levantamento de fauna.

1. Lista de espécies encontradas, indicando a forma de registro e habitat, destacando as espécies ameaçadas de extinção, as endêmicas, as consideradas raras, as não descritas previamente para a área estudada ou pela ciência, as passíveis de serem utilizadas, como indicadoras de qualidade ambiental, as de importância econômica e cinegética, as potencialmente invasoras ou de risco epidemiológico, inclusive domésticas, e as migratórias e suas rotas;

2. Caracterização do ambiente encontrado na área de influência do empreendimento, com descrição dos tipos de habitats encontrados (incluindo áreas antropizadas como pastagens, plantações e outras áreas manejadas). Os tipos de habitats deverão ser mapeados, com indicação dos seus tamanhos em termos percentuais e absolutos, além de indicar os pontos amostrados para cada grupo taxonômico;

3. Esforço e eficiência amostral, parâmetros de riqueza e abundância das espécies, índice de diversidade e demais análises estatísticas pertinentes, por fitofisionomia e grupo inventariado, contemplando a sazonalidade em cada área amostrada;

4. Anexo digital com lista dos dados brutos dos registros de todos os espécimes – forma de registro, local georreferenciado, habitat e data;

5. Detalhamento da captura, tipo de marcação, triagem e dos demais procedimentos a serem adotados para os exemplares capturados ou coletados (Vivos ou mortos), informando o tipo de identificação individual, registro e biometria.

6. Os resultados do levantamento de Ictiofauna e Invertebrados Aquáticos deverão ser incluídos os seguintes itens:
   a. Determinação dos parâmetros físico-químicos dos cursos d'água, conforme disposto na Resolução CONAMA nº357, de 2005;
   b. Parâmetros ecológicos de riqueza e abundância de espécies, bem como índice de diversidade para as comunidades de peixes, ictioplâncton, fitoplâncton e zooplâncton que deverão ser inventariadas sazonalmente, em todos os ambientes aquáticos;
   c. Destino dos exemplares capturados;
   d. Informação do lote, destino e composição quali-quantitativa de espécies em cada montante de espécimes translocados em um recipiente;

- D2 -

REQUEST FOR EVALUATION OF RESEARCH REPORT

1. **Program area:** Environment

2. **Name of study program:** Master in Environmental Management and Audits

3. **Personal details:**

Student name: Solange Batista Damasceno

Date of birth: July 6, 1972

Place of birth: Manaus, Amazona, Brazil

Student login: Solange Batista Damasceno

E-mail: bio.solange@yahoo.com.br

Date of enrollment: January 25, 2013

4. **Advisor:** Dr. Erik Simoes

5. **Title of the Research Memo:** Fauna Diversity in the Miriti Environmental Protection Area, Municipality of Manacapuru, Amazonas, Brazil

6. **Summary:**

6.1. Summary

The diversity of fauna in the Miriti Environmental Protection Area was 163 species from 80 different families, of which 12 species are classified as endangered. Of the species, 36% are Avifauna, 33% Ichthyofauna, 15% Mastofauna and 16% Herpetofauna. Of the families, 35% are Avifauna, 22% Ichthyofauna, 24% Mastofauna and 19% Herpetofauna.

6.2. Abstract

The diversity of species in the Environmental Protection Area of Miriti was cataloged 163 species of 80 different families, of which 12 species are classified as endangered. Species, are 36% of bird life, 33% Ichthyofauna 15% of Mastofauna and 16% of Herpetofauna. Families are 35% of bird life, 22% Ichthyofauna, 24% of Mammalogy, 19% of reptiles and amphibians.

7. Key words:

Present the five key words that represent the central theme of your research. They can be the same as those used in the Research Memoir.

Diversity; Fauna; APA; Cataloging; Miriti.

8. Curriculum vitae:

Only fill in the spaces that correspond and with information from the last three years. Do not indicate more than three points per topic.

Academic background:

Degree in Biological Sciences

Further training:

Specialization in Sustainable Development

Professional Experience/Occupation:

Water Resources Management

Teaching experience:

Postgraduate - Modular

Research experience:

Patents

No

R&D&I projects

No

Publications

Booklet - Taruma-Açu doing well by the water

Other merits of interest

Environmental Consulting and Advisory (Wildlife)

9. Student regularity:

By means of this document, Solange Batista Damasceno, a student in the Environment and Sustainable Development academic program, expresses her agreement with the Research Report submitted and formally requests that it be evaluated.

Brazil, November 7, 2016.

Assinatura: _Solange Batista Damasceno._

(inserir a imagem de sua assinatura digitalizada)

10. Resolution (to be completed by the Evaluating Court):

Evaluation result: Select the option.

Final Score: Average of the scores of the members of the judging panel.

Comments:

Write all your comments in this space

Buy your books fast and straightforward online - at one of world's fastest growing online book stores! Environmentally sound due to Print-on-Demand technologies.

Buy your books online at
www.morebooks.shop

Kaufen Sie Ihre Bücher schnell und unkompliziert online – auf einer der am schnellsten wachsenden Buchhandelsplattformen weltweit! Dank Print-On-Demand umwelt- und ressourcenschonend produziert.

Bücher schneller online kaufen
www.morebooks.shop

Printed by Books on Demand GmbH, Norderstedt / Germany